A BASIC GUIDE TO SMALL ENGINE REPAIR

Jason Francis

A Basic Guide to Small Engine Repair

DISCLAIMER

The information provided in this book is for educational and informational purposes only. The author and publisher make no representations or warranties regarding the accuracy, completeness, or usefulness of the information contained in this book. The author and publisher shall not be held liable for any damage, injuries, or losses incurred as a result of the use of the information provided in this book.

ACKNOWLEDGMENTS OF CONTENT

The author and publisher acknowledge the use of third-party materials in this book, including but not limited to images, quotes, or excerpts, and have made efforts to ensure proper attribution is given where applicable. Any errors or omissions are unintentional, and the publisher will make efforts to rectify them in future editions.

Cover Design By: Jason Francis
Edited By: Kelsey Hilborn McCoy
ISBN: 979-8-218-99298-9 (paperback)
ISBN: 979-8-218-99299-6 (ebook)

DEDICATION

To my family, whose love and support have been the foundation of everything I do. You have always been there for me, provided encouragement, and reminded me of what truly matters. Each of you has played a role in shaping who I am today, and for that, I am eternally grateful.

To my friends, for sharing your laughter, wisdom, and countless memories. Your companionship has brought joy and balance to my life, and you've been a constant source of inspiration and strength. The moments we've shared will forever hold a special place in my heart.

To my co-workers, with whom I have had the privilege of collaborating and learning. Your expertise, dedication, and camaraderie have made each challenge easier to face and every success sweeter to celebrate. I treasure our teamwork and mutual respect.

Most importantly, to my incredible wife. You are my rock, my partner, and my greatest source of love and inspiration. Your patience, encouragement, and belief in me have been my constant guide, making every step of this journey worthwhile. This book is as much yours as it is mine. Thank you for being by my side every step of the way.

ACKNOWLEDGMENTS

I would like to express my sincere gratitude to the small engine repair program in high school, which provided me with invaluable skills and hands-on experience. A special thanks to my instructor, who patiently guided me through the complexities of engine mechanics, fostering both my curiosity and technical knowledge. Your dedication to teaching and your ability to simplify challenging concepts made learning not only enjoyable but also deeply rewarding.

I also want to thank my classmates for their camaraderie and teamwork. Together, we tackled projects, shared insights, and supported each other in ways that made the learning process both fun and effective.

Finally, I'm grateful to my family for encouraging me to pursue this path and for their continuous support throughout my journey. The skills I gained in this program have shaped my understanding of mechanics and instilled in me a lasting passion for working with engines.

TABLE OF CONTENTS

INTRODUCTION

As small engine repair costs continue to rise, consumers face increased amounts of expensive repair bills to have their equipment operate in tip-top shape. With the growing demand for small engine services, local repair shops have raised their rates, leaving individuals looking for ways to address common issues without breaking the bank. This book is designed to empower you to troubleshoot, diagnose, and repair small engine problems on your own, providing practical, step-by-step instructions to help you get your equipment up and operating again.

Small engine repairs often seem complicated, and individuals assume they need a professional to fix their issues. However, the reality is that problems with lawn mowers, chain saws, generators, and other small engine-powered equipment can be managed with little mechanical knowledge and the right set of tools. This book aims to provide you with both. It's not just a guide, it is a comprehensive resource that will equip you with the knowledge, skills, and confidence to take control of your small engine maintenance and repair needs.

Whether you're dealing with a lawn mower that refuses to start, a chain saw that's lost its cutting power, or an old generator that's in desperate need of a tune-up, this book will guide you through the troubleshooting process, helping you to identify the root of the problem and walking you through the steps to fix it. The key to addressing these common issues is understanding the basics of how small engines work and gaining the right knowledge to make informed decisions about maintenance and repairs.

The goal of this book is to break down the barriers that may have once made small engine repairs feel intimidating. Each chapter includes practical solutions and helpful tips, tricks, and reminders that are designed to be easily followed, especially for beginners. Small engine repair doesn't need to be a mystery, and with the clear instructions provided in this guide, even those with no prior experience can begin to develop the skills needed to tackle small engine issues.

Throughout this book, there will be a wide range of common small engine problems covered, including engine starting issues, fuel system malfunctions, air filter blockages, spark plug failures, and much more. For each issue, potential solutions will be covered first, followed by a set of straightforward steps for diagnosing and fixing the problem. From basic engine maintenance, such as oil changes and spark plug replacements, to more advanced tasks, such as carburetor cleaning and valve adjustments, this guide will provide you with the tools to tackle both simple and complex repairs.

In addition to fixing problems, this book also emphasizes preventive maintenance. Regular upkeep is one of the most effective ways to avoid costly repairs down the line. You will learn how to perform routine maintenance tasks, such as cleaning air filters, checking oil levels, and adjusting the carburetor, to ensure that your equipment continues to run smoothly. By understanding the inner workings of a small engine and how to maintain it, you will reduce the likelihood of breakdowns and extend the lifespan of your engines.

The chapters are organized in a way that will build your understanding of small engine systems. The book starts with foundational knowledge, such as engine parts and how they function together, then moves on to step-by-step instructions for troubleshooting common problems. Each chapter is written with a focus on clarity and simplicity, ensuring that even first-time repairers can follow along. There isn't any technical jargon or complicated terms in the book, just practical, easy-to-understand guidance that helps you get the job done.

As you work through the chapters, you will find that small engine repairs are more about problem-solving than they are about performing technical procedures. The key is understanding the symptoms, identifying possible causes, and then applying the right fix. This book walks you through the thought process, encouraging you to diagnose and repair your engine with confidence. Whether you are looking to make minor adjustments or are ready to dive into more complex repairs, you will always have a roadmap to follow.

The act of repairing something yourself is not only empowering but also rewarding. It gives you a sense of accomplishment and helps you develop a deeper appreciation for the machines you rely on daily. By learning how to repair and maintain your own small engines, you will gain a deeper understanding of how they work, which will make you a smarter consumer over time. You will know when a problem requires professional assistance and when it's something you can handle on your own, saving you time, money, and stress.

For those of you who are considering pursuing a career in small engine repair, this book also provides a solid foundation upon which you can build further expertise.

The skills and knowledge outlined in these pages will help you begin your journey in small engine repair, whether as a hobbyist, a professional mechanic, or someone simply looking to save money by managing your own equipment maintenance. If you find yourself enjoying the process of repairing small engines, there are opportunities to further your knowledge and experience in this field, including enrolling in professional certification programs or working in a local repair shop.

The beauty of small engine repair is that it is accessible to anyone. You don't need a fancy workshop or expensive tools to get started. Most of the necessary tools can be found at your local hardware store, and repairs can be completed with just a few basic tools, such as screwdrivers, wrenches, pliers, and a spark plug socket. The key is learning how to use these tools and understanding the purpose of each component of your small engine. As you become more experienced, you will find that diagnosing and fixing problems become second nature. One of the greatest advantages of learning small engine repair is that it helps you avoid unnecessary repair shop fees. As labor costs rise, even simple repairs can become expensive when taken to a mechanic. By learning how to do the work yourself, you can save money and keep your machines running smoothly without the additional expense of hiring professionals. Overall, the cost savings can be significant, and you will have the satisfaction of knowing that you fixed the problem yourself.

At its core, this book is about empowerment. It is about providing you with knowledge and tools to take control of your small engine maintenance and repairs. Rather than relying on others, you will be able to manage the most common issues with confidence, saving money and time along the way. Whether you are looking to perform basic maintenance or tackle more advanced repairs, this guide is your go-to resource for small engine care.

As you finish reading this book, you will not only have the skills to repair and maintain your equipment but also the confidence to take on any small engine challenge that comes your way. From diagnosing problems to making repairs and performing regular maintenance, you'll have all the tools and knowledge needed to keep your equipment running at its best.

OVERVIEW OF THE BOOK

Welcome to "A Basic Guide to Small Engine Repair," a comprehensive resource designed to help you understand, maintain, and repair small engines. Whether you're a homeowner trying to keep your lawn mower in top shape, a DIY enthusiast looking to learn more about engine mechanics, or a professional technician seeking to deepen your knowledge, this book has something for everyone. Small engines power a wide range of tools and machines that make our lives easier, from lawn mowers and chain saws to generators, snowblowers, and even go-karts. With the knowledge contained in this book, you will have the skills and confidence to tackle a variety of engine-related issues on your own.

Small engines are known for their compact size, efficiency, and affordability. Despite their size, these engines are intricate machines that require regular maintenance and occasional repairs to function properly. Many common small engine problems are relatively simple to diagnose and fix, provided you have the right knowledge and tools. The objective of this book is to equip you with the practical skills to maintain, troubleshoot, and repair small engines, saving you time and money on repairs while deepening your understanding of engine mechanics.

This book is divided into four sections. The chapters within each part build upon one another, starting with foundational knowledge and progressing to more advanced repair techniques. By the end of this guide, you will be well prepared to perform small engine repairs with confidence, whether you are a beginner or have some experience.

Part 1: Introduction to the Small Engine

Chapter 1: Introduction to Small Engines

The first step in understanding small engine repair is to familiarize yourself with the engine itself. This chapter introduces you to the different types of small engines commonly used in tools such as lawn mowers, chain saws, and generators. Key components such as the piston, crankshaft, spark plug, carburetor, and valves are discussed in detail, highlighting their specific roles in generating power. Understanding these parts is essential for diagnosing problems and performing repairs.

Although small engines are designed to be durable, they require regular maintenance and care. This chapter sets the foundation for a deeper exploration of the engine's inner workings and the maintenance tasks that keep it running smoothly.

Chapter 2: Understanding How Small Engines Work

In this chapter, you will learn how small engine components work together to produce power. Small engines typically operate on either a two-cycle or four-cycle system, also known as the "stroke." The chapter compares these two types of engines, explaining their differences in terms of the power cycle. A two-cycle engine completes a power cycle in just two strokes of the piston, while a four-cycle engine requires four strokes: intake, compression, power, and exhaust. Understanding the four stages of the four-cycle engine in detail will help you troubleshoot performance issues. Each stage, including air and fuel intake, compression, ignition, and exhaust, are vital for the engine's operation. This knowledge will aid in diagnosing any issues related to the performance of the engine.

Chapter 3: Essential Tools and Safety Practices

Before tackling engine repairs, it is important to have the right tools and safety practices in place. This chapter discusses the essential tools required for small engine repair, including screwdrivers, wrenches, diagnostic tools, and specialized equipment such as carburetor cleaning kits and valve adjustment tools. Additionally, proper safety precautions are emphasized. Using personal protective equipment (PPE) such as gloves, goggles, and hearing protection is critical when working with small engines. This chapter provides safety tips for handling fuel, managing fire risks, and preventing accidents, ensuring you remain safe while working on your engine.

Chapter 4: Resources and Further Learning

If you want to expand your skills beyond this book, this chapter provides valuable resources. From online tutorials and repair guides to community forums and local workshops, many tools and platforms are available to deepen your knowledge of small engine repair. Practical experience is key, and engaging with the repair community can boost your confidence and expertise over time.

Part 2: Maintenance and Troubleshooting

Chapter 5: Maintenance Basics

Routine maintenance is essential for ensuring that your small engine operates at peak efficiency. This chapter covers basic maintenance tasks such as oil changes, air filter replacement, spark plug inspection, and carburetor cleaning. These simple tasks prevent major issues and keep your engine running smoothly. Additionally, proper storage techniques during the offseason are discussed, ensuring your engine is protected from weather elements and idle wear. Regular maintenance doesn't have to be complicated, and by following these simple tasks, you can save money on costly repairs.

Chapter 6: Troubleshooting Common Problems

When a small engine starts to malfunction, the first step is to identify the root cause. Chapter 6 provides a guide to troubleshoot common problems, including failure to start, rough idling, stalling, and decreased power. It explains how to systematically check components such as the fuel system, spark plug, and air filter. A detailed diagnostic checklist helps you work through the troubleshooting process to ensure that no part is overlooked. This methodical approach allows you to resolve issues quickly and efficiently.

Chapter 7: Troubleshooting FAQs

Some engine problems can be more difficult to diagnose, and this chapter addresses frequently asked questions that may arise during repairs. It offers troubleshooting tips for diagnosing starting difficulties, unusual noises, and performance issues. This FAQ section serves as a quick reference for resolving common issues without unnecessary repairs or a mechanic's visit.

Chapter 8: Fuel System Repair

The fuel system is critical to the performance of any engine. Chapter 8 covers the diagnosis and repair of common fuel system problems, including issues with the carburetor, fuel lines, and fuel filter. You'll learn how to clean and rebuild a carburetor, fix fuel delivery issues, and replace clogged or worn fuel filters. Carburetors are particularly prone to clogging because of dirt or old fuel. This chapter provides step-by-step instructions for disassembling, cleaning, and reassembling a carburetor, ensuring the engine receives optimal fuel flow.

Chapter 9: Electrical System Repair

The electrical system of a small engine includes components such as the ignition coil, spark plug, and wiring. In Chapter 9, you'll learn how to troubleshoot and repair electrical system issues. The chapter walks you through testing the ignition system, diagnosing faulty components, and replacing broken parts. Electrical problems can sometimes be difficult to troubleshoot, but with the right tools and knowledge, you will be able to identify and fix issues such as faulty spark plugs, broken wires, or defective ignition coils. Testing the electrical system with a multimeter is also covered in this chapter.

Chapter 10: Mechanical Repairs

Some small engine repairs involve internal components such as pistons, valves, camshafts, and gaskets. Chapter 10 explores how to diagnose and repair mechanical issues such as low compression, oil leaks, or engine knocking. You will learn how to disassemble and reassemble the engine, replace worn parts, and address internal engine problems. Performing mechanical repairs can be challenging, but with the information provided in this chapter, you'll gain the skills needed to tackle more complex issues.

Part 3: Rebuilding the Small Engine

Chapter 11: Small Engine Rebuilds

Rebuilding a small engine is a rewarding but intricate process. This chapter guides you through disassembling the engine, inspecting parts for wear, replacing components, and reassembling the engine. Rebuilding can be a cost-effective way to extend the life of an engine and avoid purchasing a new piece of equipment. You'll learn how to identify when a rebuild is necessary, how to source replacement parts, and how to perform the rebuild step by step.

Part 4: Prevention

Chapter 12: Preventing Future Issues

The final chapter discusses strategies for preventing engine problems before they arise. By following a regular maintenance schedule and adopting proactive care measures, you can prevent many of the most common engine issues. Environmental considerations, such as properly disposing of old fuel, fluids, and engine parts, are also addressed. This practice helps minimize waste and reduces the environmental impact. Consistent care and attention to your small engine will ensure it runs smoothly for years to come.

PART 1

INTRODUCTION TO THE SMALL ENGINE

CHAPTER 1

INTRODUCTION TO SMALL ENGINES

Have you ever been in the middle of a project, where your hands are covered in dirt or oil, when your small engine suddenly sputters and dies for no apparent reason? You try to restart it, but it refuses to come back to life. The frustration builds, and the work halts. If this scenario sounds familiar, you are not alone. Small engine failures often occur at the most inconvenient times, leaving you feeling helpless and unproductive. Fortunately, this book will help you understand, troubleshoot, and repair small engines, offering practical insights and solutions that will keep your equipment running smoothly when you need it most.

Small engines power a variety of machines across different industries, from landscaping and construction to emergency power generation. These compact, yet powerful, engines are critical for the operation of everyday equipment. Yet, despite their importance, a lot of people are unfamiliar with how small engines operate and how to properly maintain or even repair them. This book aims to bridge that knowledge gap. By understanding the mechanics behind small engines and learning how to troubleshoot common issues, you'll be able to address problems quickly and confidently, reducing downtime and minimizing frustration. This knowledge will not only save you time and money but will also empower you to take control of your equipment and avoid costly repairs. Regular maintenance is essential for keeping small engines running efficiently, and this book provides you with step-by-step guidance on how to perform routine maintenance, replace worn-out parts, and resolve issues that may arise.

The applications of small engines span a broad range of industries. For example, in landscaping, small engines are the backbone of equipment used to maintain residential and commercial properties. Lawn mowers, leaf blowers, trimmers, and hedge trimmers all rely on small engines to function. In the construction industry, small engines power tools such as concrete mixers, compactors, and generators, enabling workers to complete tasks more efficiently. Small engines also play a vital role in emergency power generation. During power outages or in remote locations where electricity is unavailable, portable generators powered by small engines provide a critical source of backup power, ensuring that essential systems, from lights to medical equipment, continue to function.

Each of these applications has unique requirements and challenges. A small engine used in a lawn mower may have different maintenance needs than one used in a chain saw or a snowblower. Understanding the specific needs of each type of engine is vital for ensuring optimal performance. For instance, lawn mower engines are often exposed to grass clippings and debris, requiring regular cleaning to prevent clogging, while snowblower engines face harsh winter conditions that can cause starting issues. Knowing how to properly care for each engine type will help you maximize its lifespan and keep it running smoothly.

In the following chapters, you will dive deeper into the mechanics of small engines, common problems that can occur, and how to fix them. You will explore how to diagnose engine failures, replace spark plugs, clean carburetors, and perform other essential maintenance tasks. By the end of this book, you will have the knowledge and confidence to tackle small engine repairs on your own, saving you both time and money while keeping your equipment in top working condition.

LAWN MOWERS: POWERING EVERYDAY LAWN CARE

Lawn mowers are the most familiar application of small engines in daily life. These engines are designed with simplicity and cost-effectiveness in mind, often ranging from 3 to 7 horsepower and featuring a single-cylinder design. This setup allows for easy maintenance and repairs, making small engines in lawn mowers an accessible option for homeowners and lawn care professionals alike. The engine in a lawn mower is responsible for driving the blade that cuts the grass, as well as powering other functions, such as self-propulsion. Lawn mowers operate at lower revolutions per minute (RPM) compared to other small engine applications, but they still need to provide consistent power for long periods to ensure efficient operation. Engine design varies depending on the type of mower; walk-behind mowers, push mowers, and riding mowers all use different engine configurations to match their intended tasks.

Key Features:

Horsepower: Typically, lawn mowers have between 3 to 7 horsepower.

Engine Design: Single-cylinder, air-cooled engines are common.

Fuel Type: Gasoline is commonly used, although electric and battery-powered mowers are growing in popularity. As one of the most widely used pieces of equipment globally, proper maintenance and understanding of the engine's operation are critical. Regular oil changes, air filter replacements, and spark plug inspections are essential to keeping lawn mowers running efficiently.

GENERATORS: RELIABLE POWER FOR REMOTE LOCATIONS AND EMERGENCIES

Small engines also power portable generators, which are indispensable in a variety of situations. These engines are used for outdoor events, such as jobsites that do not have access to electrical power, or are used as backup power during outages.

These engines must be both dependable and durable, capable of running for extended periods without failure, which makes them a unique challenge to maintain and repair.

Typically ranging from 5 to 20 horsepower, the engines in portable generators are designed to run continuously and produce stable, consistent power for hours on end. Depending on the size of the generator, the engine will either be air-cooled or liquid-cooled to prevent overheating during extended use. Smaller generators, particularly those used for camping or light-duty applications, often feature a simple single-cylinder engine that balances portability and efficiency.

Key Features:

Horsepower: Typically, generators range from 5 to 20 horsepower, depending on the generator size.

Cooling System: Generators can be air-cooled or liquid-cooled, depending on the engine's size and application.

Fuel Type: Gasoline, diesel, natural gas, and propane are common fuel types, with gasoline being the most widely used. Generators often require less frequent maintenance than lawn mowers but still need regular checks on oil levels, fuel systems, and electrical components to ensure they operate smoothly when needed most.

CHAIN SAWS: HIGH-PERFORMANCE POWER FOR TOUGH JOBS

Chain saws are another prominent application of small engines, particularly in the forestry, landscaping, and construction industries. These tools are designed for cutting wood, clearing forests, and performing heavy-duty tasks that require both power and portability. Small engines in chain saws typically range from 2 to 6 horsepower and are engineered for lightweight portability without sacrificing performance. Chain saw engines are built for high RPMs, allowing the saw to cut through wood quickly and efficiently. The engine must provide the power needed to keep the blade spinning at optimal speeds while also managing the vibrations associated with cutting. This ability requires a finely tuned balance of torque, horsepower, and engine design to maximize cutting performance while minimizing user fatigue.

Key Features:

Horsepower: Typically, chain saws have 2 to 6 horsepower.

Engine Design: Chain saws often have a two-stroke engine, which offers more power for its size and allows the chain saw to be lightweight and portable.

Fuel Type: Typically, chain saws run on a gasoline and oil mixture in a two-stroke engine. Maintaining a chain saw engine involves checking the fuel lines, fuel filter, air filters, spark plugs, and carburetor. It is also important to regularly clean the chain and bar, as well as ensure the engine is properly lubricated, to keep it in top working condition.

SNOWBLOWERS, PRESSURE WASHERS, AND TRIMMERS: SPECIALIZED TASKS IN VARIOUS ENVIRONMENTS

Small engines also power a wide range of other specialized equipment used in both residential and commercial applications. Snowblowers, pressure washers, and string trimmers are all tools that rely on small engines to provide the necessary power to tackle specific tasks, such as clearing snow, washing surfaces, or trimming grass and weeds.

Snowblowers: These machines use small engines to propel the blades and move snow. Typically found in residential areas where snow accumulation is common, snowblowers can range from 5 to 13 horsepower, with larger models offering more power for clearing deeper snow. The engine must operate reliably in cold temperatures and be capable of moving large amounts of snow quickly.

Pressure Washers: Used for cleaning surfaces such as driveways, siding, and vehicles, pressure washers use small engines to power high-pressure water pumps. These engines typically range from 3 to 10 horsepower and can vary in complexity depending on the model, with commercial-grade washers requiring more robust engines.

String Trimmers (Weed Eaters): Used for trimming grass and weeds in areas that are difficult to reach with a lawn mower, these devices typically feature engines in the 1 to 3 horsepower range. Trimmers often use two-stroke engines, which are lightweight and efficient for the task.

Key Features:

Horsepower: Snowblowers range from 5 to 13 horsepower, pressure washers range from 3 to 10 horsepower, and trimmers usually range from 1 to 3 horsepower.

Fuel Type: Gasoline is common for these devices, though some models may be electric or battery-operated.

Portability: Most of these engines are designed to be lightweight and easy to transport, especially in residential applications.

Maintenance for these types of equipment includes cleaning air filters, checking the fuel system, lubricating moving parts, and ensuring the engine is properly fueled and primed before use.

OTHER APPLICATIONS FOR SMALL ENGINES

Beyond the more common applications already discussed, small engines are also found in a wide range of other equipment. Some additional notable examples include:

Go-Karts: Small engines are used in go-karts for recreational and racing purposes. These engines typically range from 5 to 15 horsepower and must be designed for high RPMs and durability.

Leaf Blowers: Used for clearing leaves and debris from yards or streets, leaf blowers rely on small, high-performance engines, typically between 1 to 5 horsepower.

Air Compressors: Small engines power portable air compressors used in a variety of applications, from inflating tires to powering pneumatic tools. These engines usually range from 3 to 7 horsepower.

Agricultural Equipment: Small engines are used in garden tillers, water pumps, and other agricultural tools to help with farming tasks. These engines typically range from 3 to 10 horsepower, providing the power needed for efficient operation in the field.

THE VERSATILITY OF SMALL ENGINES

Small engines are indispensable in different aspects of modern life, powering everything from lawn mowers to generators, chain saws, and specialized equipment such as snowblowers and pressure washers. Their ability to provide efficient, reliable power in a compact form makes them ideal for a wide range of applications across residential, commercial, and industrial sectors. Understanding how small engines work and the role they play in everyday tasks is essential for anyone involved in the maintenance, repair, or operation of this equipment. Whether you're using a lawn mower to maintain your yard, powering a generator during an emergency, or cutting firewood with a chain saw, small engines are the heart of tools that help make life more manageable and productive.

CHAPTER 2

UNDERSTANDING HOW SMALL ENGINES WORK

Small engines power a plethora of tools and machines that are essential in both our daily lives and work environments. While they are simpler than the engines found in larger vehicles, small engines still rely on the same basic principles of internal combustion. This section will break down the mechanics of how small engines work, focusing on both the four-stroke and two-stroke engine, the differences between the two, and the key components that make small engines run efficiently.

THE FOUR-STROKE ENGINE

The four-stroke engine is the most common engine used in small engines, including those found in lawn mowers, generators, and many other pieces of outdoor equipment. It is called "four-stroke" because it takes four distinct cycles, or strokes, of the piston to complete one full cycle. These strokes occur in a specific order, and each plays a critical role in converting fuel into mechanical energy. The four strokes are:

1. Intake Stroke

During the intake stroke, the intake valve opens, and the piston moves downward. As the piston moves down the cylinder, it creates a vacuum that draws in a mixture of fuel and air from the carburetor into the combustion chamber. This mixture is critical for combustion because it provides the energy needed to power the engine. As the intake valve closes, the piston reaches the bottom of the stroke, trapping the fuel-air mixture inside the cylinder.

2. Compression Stroke

In the compression stroke, the piston begins to move upward, compressing the fuel-air mixture inside the cylinder. As the piston moves up, both the intake and exhaust valves remain closed, ensuring that the air-fuel mixture is compressed. This process increases the pressure and temperature of the mixture, making it more combustible. The compression stroke is vital because a more compressed mixture ignites more efficiently, generating more power.

3. Power Stroke

Once the fuel-air mixture has been compressed, the spark plug ignites it. The spark from the spark plug causes the mixture to combust (explode), and this explosion forces the piston downward. This is the stroke that generates the engine's power. The power stroke is what ultimately drives the engine's flywheel and turns the crankshaft, providing the mechanical energy that powers the equipment. It is the most forceful stroke in the cycle and is responsible for doing much of the work.

4. Exhaust Stroke

Finally, in the exhaust stroke, the exhaust valve opens, and the piston moves upward again. As the piston moves up, it pushes the exhaust gases (the byproducts of combustion) out of the cylinder through the exhaust valve. Once the piston reaches the top of its stroke, the exhaust valve closes, and the intake valve opens, starting the cycle over again with the intake of a new fuel-air mixture. This exhaust stroke ensures that the engine runs cleanly by removing the waste gases generated during combustion.

The four-stroke, or four-cycle, is highly efficient and provides smooth, reliable power. It also ensures that the engine produces less smoke and emissions compared to older engine types, which is why it's commonly used in small engines today.

DIFFERENCES BETWEEN TWO-STROKE AND FOUR-STROKE ENGINES

While the four-stroke engine is the most common among small engines, there are also two-stroke engines, which are often found in smaller, lighter equipment such as chain saws, weed trimmers, and some motorcycles. Understanding the differences between two-stroke and four-stroke engines can help clarify why one may be chosen over the other for specific applications.

TWO-STROKE ENGINES

In a two-stroke engine, the piston completes a power cycle in just two strokes (one up and one down), rather than the four strokes required in a four-stroke engine. This means the two-stroke engine has a simpler design, with fewer components and parts to maintain.

HOW THE TWO-STROKE ENGINE WORKS

Intake and Compression: As the piston moves upward, it compresses the fuel-air mixture while simultaneously drawing in a new charge of fuel and air into the crankcase. The intake and exhaust ports are uncovered by the piston, allowing the new mixture to enter the crankcase.

Power and Exhaust: When the piston reaches the top, the spark plug ignites the mixture. The explosion pushes the piston downward, providing power. As the piston moves down, it simultaneously pushes the exhaust gases out of the exhaust port and the new fuel-air mixture into the combustion chamber.

While two-stroke engines are lighter and more compact than four-stroke engines, they are less fuel-efficient, produce more emissions, and require a mixture of oil and fuel to lubricate the internal components. This setup makes them less environmentally friendly and harder to maintain than four-stroke engines.

FOUR-STROKE ENGINES

As described earlier, four-stroke engines are more complex, with separate strokes for intake, compression, power, and exhaust. This situation makes them more fuel-efficient because they do not require oil mixed with the fuel. Four-stroke engines tend to produce less pollution and run longer before requiring maintenance, making them ideal for applications where longevity and efficiency are critical. However, they are heavier and more expensive to produce than two-stroke engines.

While two-stroke engines are lightweight and compact, making them ideal for portable equipment such as chain saws, four-stroke engines are more efficient and cleaner, making them better suited for tasks that require longer running times and less maintenance.

KEY COMPONENTS OF SMALL ENGINES

Several key components work together to make small engines run smoothly. Understanding these components and their functions is essential for anyone interested in maintaining or repairing small engines.

Piston

The piston is a fundamental component of the small engine, playing an important role in converting fuel into usable mechanical energy. Typically made of high-strength aluminum or steel alloys, the piston is a cylindrical piece of metal that fits snugly within the engine's cylinder. As the engine operates, the piston moves up and down inside the cylinder, driven by the pressure changes created by the combustion process. This up-and-down motion is essential for the engine's functioning because it allows for the compression of the fuel-air mixture and the conversion of the resulting combustion energy into mechanical energy.

The piston is involved in several key stages of the engine's operation, most notably during the four-stroke (intake, compression, power, and exhaust). During the intake stroke, the piston moves down the cylinder, creating a vacuum that allows the intake valve to open and draw in a fuel-air mixture. The mixture fills the combustion chamber, setting the stage for the next phase.

As the piston begins to move upward during the compression stroke, it compresses the fuel-air mixture into a smaller space within the combustion chamber. This step is a critical process because the higher the compression, the more energy can be extracted from the fuel during combustion.

Once the piston reaches the top of its stroke, the spark plug ignites the compressed fuel-air mixture. The resulting explosion creates a rapid expansion of gases, which pushes the piston down during the power stroke. This step is where the piston's role becomes particularly important; the force generated by the expanding gases is what drives the piston down the cylinder. The energy from the combustion process is thus converted into the linear motion of the piston. The piston is connected to the crankshaft by the connecting rod, a strong metal rod that transmits the piston's linear motion to the crankshaft. As the piston moves down during the power stroke, it pushes the connecting rod, which in turn rotates the crankshaft. The crankshaft's rotation is then used to drive the engine's flywheel, which stores rotational energy and helps maintain a smooth, continuous operation by smoothing out the pulses of power produced during each stroke of the piston. This rotation is ultimately what powers the vehicle or equipment the engine is driving.

The piston's movement and function are essential to an engine's efficiency and performance. Any damage or wear to the piston, such as scoring or cracking, can reduce the engine's power output, efficiency, and longevity. Pistons also endure extreme temperatures and pressures, so they must be designed to withstand these stresses. To facilitate this situation, pistons typically have a layer of coating or treatment that helps reduce friction between the piston and cylinder walls, ensuring smooth movement and reducing wear. Additionally, pistons are often equipped with piston rings, which form a seal between the piston and the cylinder wall. These rings help control oil consumption, prevent combustion gases from escaping, and maintain compression. Over time, piston rings can wear out and need to be replaced to ensure the engine continues to operate efficiently.

In summary, the piston is a critical component that directly impacts an engine's performance, efficiency, and longevity. It serves the essential function of compressing the fuel-air mixture, converting combustion energy into mechanical motion, and transmitting that energy through the connecting rod to the crankshaft, where it can be used to power the engine's flywheel. Without the piston, the engine would be unable to harness the power needed for operation, making it one of the most vital parts of any internal combustion engine.

Carburetor

The carburetor is a critical component in an internal combustion engine, responsible for mixing air and fuel in the proper ratio before it enters the combustion chamber. This mixture is essential for the engine to run efficiently because the right balance of air and fuel ensures optimal combustion, power output, and fuel economy. The carburetor adjusts the air-fuel ratio dynamically, based on factors such as engine speed, load, and environmental conditions, allowing the engine to perform smoothly under various operating scenarios.

As the engine operates, the carburetor continually adjusts to ensure the engine receives the correct mixture of air and fuel. It completes this process through various mechanisms, including throttle valves, jets, and choke, which regulate the flow of air and fuel. These adjustments help maintain a stable idle, prevent stalling, and ensure the engine accelerates smoothly.

If the carburetor malfunctions, it can lead to several issues. For example, a clogged or damaged carburetor may restrict fuel flow, causing the engine to run lean (too much air, not enough fuel), resulting in poor performance, difficulty starting, or even stalling.

Alternatively, an overly rich mixture (too much fuel) can waste fuel and cause the engine to run inefficiently. Regular maintenance, such as cleaning and adjusting the carburetor, is vital to ensuring the engine runs smoothly and efficiently.

Spark Plug

The spark plug is a critical component of an internal combustion engine, playing the essential role of igniting the compressed fuel-air mixture inside the combustion chamber. It functions by creating a small, high-voltage spark that ignites the fuel mixture at precisely the right moment during the engine's power stroke. This ignition is what drives the piston downward, producing the mechanical energy needed to power the engine.

Spark plugs are designed to withstand extreme temperatures and pressures because they are exposed directly to the combustion process. The spark plug consists of a central electrode and a ground electrode, with a gap between them. When the engine's ignition system sends an electrical current through the spark plug, it creates a spark that ignites the fuel-air mixture. The timing and quality of this spark are critical for engine performance.

A faulty spark plug can lead to a range of issues, including engine misfires, rough idling, decreased power output, or a complete failure to start. It may also cause inefficient combustion, resulting in poor fuel economy and increased emissions. Regular maintenance, such as inspecting and replacing spark plugs, is essential to ensure optimal engine performance and reliability.

Crankshaft

The crankshaft is a vital component in an engine, responsible for converting the reciprocating motion of the piston into rotational motion. As the piston moves up and down inside the cylinder, the connecting rod attached to it transfers this linear movement to the crankshaft. The crankshaft then rotates, transferring the energy from the piston to the engine's flywheel, which ultimately drives the equipment connected to the engine, such as a lawn mower blade, generator, or other tools. This rotational motion is essential for the continuous operation of the engine and the machinery it powers. Made from strong, durable materials such as steel or cast iron, the crankshaft must withstand significant forces and stress during operation. It needs to resist wear and deformation over time while maintaining precision in its movement. Because the crankshaft is subjected to high pressures and rotational forces, any damage or wear to it can result in poor engine performance or even engine failure. For this reason, the durability and integrity of the crankshaft are critical for the engine's longevity. Regular maintenance, such as checking for oil levels and using the correct type of lubricant, is essential to ensure the crankshaft operates smoothly and efficiently throughout the engine's life.

Valves (Intake and Exhaust)

The intake and exhaust valves are essential components in the engine's operation, controlling the flow of gases into and out of the combustion chamber. These valves play a key role in maintaining the proper sequence of events in the engine's four-stroke, which includes the intake, compression, power, and exhaust strokes.

The intake valve opens during the intake stroke, allowing the fuel-air mixture to enter the combustion chamber. This mixture is drawn in as the piston moves down, creating a vacuum that pulls in the necessary fuel and air for the combustion process. Once the piston reaches the bottom of its stroke, the intake valve closes to seal the chamber in preparation for compression.

The exhaust valve, on the other hand, opens during the exhaust stroke. As the piston moves up, it forces the spent combustion gases out of the cylinder through the open exhaust valve, making room for a fresh charge of fuel and air. The exhaust valve closes at the top of the stroke to prevent gases from re-entering the combustion chamber during the intake stroke.

In engines equipped with overhead valves (OHV), the intake and exhaust valves are positioned above the cylinder head, with a camshaft located in the engine's block or overhead. OHV engines generally provide better fuel efficiency and improved engine performance because of more precise valve timing. Proper valve timing – coordinating the opening and closing of the valves with the piston's movement – is critical for engine efficiency. It ensures that the maximum amount of air-fuel mixture enters the combustion chamber and that exhaust gases are expelled effectively, leading to optimal engine performance, power output, and fuel economy. Misalignment or improper timing of the valves can result in inefficient combustion, loss of power, and engine damage.

Flywheel

The flywheel is a critical component in any small engine, playing several vital roles in ensuring smooth and consistent operation. Positioned on the engine's crankshaft, the flywheel stores rotational energy generated by the engine's power strokes. As the piston moves through the power stroke, the flywheel absorbs and retains some of the rotational energy, which helps to smooth out the engine's power delivery. This action reduces the fluctuations in rotational speed, providing a more consistent and stable power output. Without the flywheel, the engine would experience uneven power surges, leading to rough performance and less efficient operation.

In addition to smoothing out the engine's power, the flywheel helps maintain a steady engine speed. It does this by resisting rapid changes in rotational speed, which is especially important during the engine's idle and deceleration phases. By resisting sudden changes, the flywheel ensures that the engine continues to run at a steady pace, even when the load varies or the throttle is adjusted.

Most flywheels also incorporate magnets, which are essential for powering the engine's ignition system. These magnets create a magnetic field that is used by the ignition coil to generate the electrical current needed to fire the spark plug. This integration of the flywheel and ignition system ensures that the engine's spark plug receives the necessary energy to ignite the fuel-air mixture in the combustion chamber, enabling the engine to start and run efficiently. The flywheel's combined roles of energy storage, power smoothing, and ignition assistance make it a critical component in the overall functioning of a small engine.

Governor

A governor on a small engine is a critical component that is designed to regulate the engine's speed and maintain optimal performance under varying conditions. Its primary function is to control the engine's RPM, adjusting the throttle to ensure that the engine operates within a safe and efficient speed range. This regulation is especially important when the load on the engine changes, such as when a lawn mower encounters thicker grass or a generator is required to power additional devices.

The governor works by sensing changes in engine speed and automatically adjusts the throttle. When the engine's RPM increases because of a lighter load, the governor reduces the throttle to bring the RPM back down to a preset level. Conversely, when the load increases and the RPM drops, the governor opens the throttle to allow more fuel and air into the engine, boosting power to maintain a consistent speed. This balance prevents the engine from over-revving or stalling, ensuring smoother operation and reducing the risk of engine damage.

Governors are commonly found in small engines used in lawn mowers, generators, pressure washers, and other power equipment. They are typically mechanical or electronic in design. Mechanical governors use centrifugal force, while electronic governors rely on sensors and actuators for more precise control. Regardless of the type, governors play a vital role in enhancing engine performance, improving fuel efficiency, and extending the engine's lifespan by preventing stress from operating outside of its optimal speed range.

Camshaft

The camshaft is a vital component in a small engine that controls the timing and operation of the intake and exhaust valves. It is a rotating shaft with strategically placed lobes that push the valves open and closed at precisely timed intervals in relation to the crankshaft's rotation. The camshaft works in synchronization with the crankshaft, ensuring that the intake and exhaust valves open and close at the correct times during the engine's four-stroke, which includes the intake, compression, power, and exhaust strokes.

During the intake stroke, the camshaft operates the intake valve, allowing the air-fuel mixture to enter the combustion chamber. During the exhaust stroke, the camshaft opens the exhaust valve, releasing the combustion gases. The precise timing of these valve movements is critical for ensuring that the engine receives the right amount of air and fuel for combustion and that the exhaust gases are efficiently expelled. This timing is fundamental for the engine's efficiency, power output, and overall performance.

The camshaft's design, including the shape and size of its lobes, can significantly impact engine characteristics, such as power delivery and fuel efficiency. For example, camshafts with larger lobes may increase power at higher RPMs, while those with smaller lobes can enhance low-end torque and fuel economy. The camshaft plays a key role in optimizing engine performance, ensuring smooth operation, and maximizing fuel efficiency while minimizing emissions.

Coil

An ignition coil is a critical component in a small engine's ignition system, responsible for generating the high-voltage spark required to ignite the fuel-air mixture in the engine's cylinder. Without this spark, the engine cannot start or operate. The ignition coil operates by transforming the low voltage from the engine's battery or magneto into the high voltage necessary to create the spark at the spark plug.

The ignition coil consists of two main parts: the primary winding and the secondary winding. The primary winding is connected to the battery or magneto, while the secondary winding is connected to the spark plug. When a current flows through the primary coil, it generates a magnetic field. As the current is suddenly interrupted (typically by the ignition points or electronic switch), the magnetic field collapses, causing a rapid rise in voltage in the secondary coil. This high voltage is then sent to the spark plug, which creates a spark to ignite the air-fuel mixture inside the combustion chamber.

The voltage generated by the ignition coil is typically in the range of 12,000 to 45,000 volts, depending on the engine's design. This powerful spark is essential for engine starting and smooth operation, ensuring efficient combustion and performance. A malfunctioning ignition coil can lead to issues such as misfires, poor engine performance, or failure to start. Thus, maintaining the ignition coil is key to keeping a small engine running reliably.

Understanding how small engines work is essential for both operators and mechanics. Whether you're troubleshooting a malfunctioning lawn mower or simply curious about the technology behind your tools, knowing the basics of the engine cycle, the differences between two-stroke and four-stroke engines, and the function of key components will help you appreciate the complexity and efficiency of these machines. With the right knowledge, you can maintain and repair small engines, ensuring they continue to provide reliable power for years to come.

CHAPTER 3

ESSENTIAL TOOLS AND SAFETY PRACTICES

Maintaining and repairing small engines requires a good understanding of both the tools you'll need and the safety practices that must be followed. Having the right equipment and a well-organized workspace can make a significant difference in the success of your tasks. Equally important is maintaining a strong focus on safety. Working with small engines can involve sharp tools, flammable materials, and potential hazards such as moving parts and hot surfaces, so understanding and adhering to safety practices is critical.

This chapter will guide you through the essential tools needed for small engine maintenance and repair, how to set up and organize your workspace, and the key safety precautions to keep in mind.

TOOLS FOR SMALL ENGINE MAINTENANCE AND REPAIR

Having the right tools on hand is critical for performing effective maintenance and repairs on small engines. Here's a list of essential tools that will help you keep your engine in top shape.

Wrenches and Socket Set

A variety of wrenches and sockets are necessary for loosening and tightening bolts, screws, and other fasteners on small engines. A combination wrench set with both standard and metric sizes will cover most tasks. A socket set with a ratchet handle and different size sockets is useful for reaching hard-to-get bolts and adjusting parts that require more torque.

Screwdrivers

A set of flathead and Phillips screwdrivers in various sizes is essential for working on smaller components such as carburetors, air filters, and ignition systems. Always use the correct size screwdriver to avoid stripping screws or damaging delicate parts.

Pliers and Needle-Nose Pliers

Pliers are used for gripping, twisting, and cutting wires, especially when working with the fuel system or electrical components. Needle-nose pliers are especially useful for reaching tight spots or manipulating small parts. Channellock pliers can be useful for larger tasks such as removing stubborn nuts or bolts.

Spark Plug Wrench

Spark plugs are integral to starting a small engine. A spark plug wrench (or socket) is designed to help you remove and replace spark plugs without damaging them. These wrenches are typically included in most basic tool kits.

Torque Wrench

Using a torque wrench ensures that bolts are tightened to the manufacturer's specifications, which is important for maintaining engine performance. Over-tightening or under-tightening can lead to leaks, improper operation, or damage to components.

Carburetor Cleaner and Brushes

Over time, the carburetor may become clogged with dirt, grime, or old fuel deposits. Using carburetor cleaner and brushes can help clean these components. A small wire brush can be helpful for removing carbon buildup on spark plugs or other metal parts.

Multimeter

A multimeter is a versatile tool used for diagnosing electrical issues, such as checking the condition of a spark plug, the battery voltage, or diagnosing problems with the ignition coil or wiring.

Fuel System Tools

Small engine fuel systems often require specialized tools such as fuel line pliers, fuel filter wrenches, and a funnel for changing fuel or checking for blockages. A vacuum pump can also help test the fuel system.

Oil and Fluid Change Tools

You'll need a drain pan to catch oil when changing it and a funnel for adding new oil. Some small engines use a special type of oil filter wrench, which is useful for removing and replacing filters when necessary.

Cleaning and Lubricating Tools

Grease guns, lubricant sprays, and cleaning cloths are essential for keeping moving parts well lubricated and free from dirt and debris. A clean engine runs more efficiently and experiences less wear and tear.

SHOP SETUP AND ORGANIZATION TIPS

Having a well-organized workspace will help you work efficiently and keep all of your tools in good condition. Here are some tips for setting up your small engine repair shop.

Choose the Right Location

Select a well-ventilated, dry, and well-lit area for your workspace. A garage or shed is usually ideal because it offers both space and shelter from the elements. Avoid working in areas where water can get onto electrical components or where flammable materials might accumulate.

Workbench

A sturdy workbench is essential for small engine repairs. Choose one with a flat surface that can support the weight of an engine and resist heavy use. Having a built-in vice or clamping tool will help secure parts while you're working on them.

Tool Storage

A proper tool storage system will help keep everything organized and accessible. Consider using tool chests or pegboards with hooks to hang frequently used tools. Organize your tools by type and size to avoid wasting time searching for the right one. Make sure that tools such as wrenches, screwdrivers, and sockets are stored in a way that prevents them from becoming lost or damaged.

Parts Storage

Small engine repairs often involve working with tiny parts such as screws, washers, and gaskets. Use small bins or drawer organizers to keep these parts separated and easy to access. Label each container for specific parts (e.g., "fuel system," "spark plugs," "fasteners," etc.).

Clean Work Area

Keep your workspace clean and free of clutter. Spills of fuel, oil, or debris can pose safety hazards and make it difficult to find tools and parts when you need them. Regularly sweep the floor, wipe down surfaces, and clean any grease or oil spills promptly.

Lighting

Ensure your workspace is lit well. It is important to note that good lighting helps reduce eye strain, allows you to see small parts more clearly, and increases overall safety. Overhead LED lighting or portable work lights can provide additional lighting, especially in corners or tight spaces.

SAFETY PRECAUTIONS

When working on small engines, safety should always be your top priority. Engine repair involves various risks, including sharp tools, flammable materials, and hot components. Following the right safety protocols can prevent accidents and injuries.

Personal Protective Equipment (PPE)

Wearing appropriate personal protective equipment (PPE) is essential when working on small engines. Here's what you should consider.

Safety Glasses: Protect your eyes from flying debris, gasoline splashes, and dust, especially when using power tools or working with the fuel system.

Work Gloves: Gloves protect your hands from sharp edges on engine components and prevent injury from handling hot parts or chemicals.

Hearing Protection: Small engines can be noisy, and extended exposure to loud sounds can lead to hearing damage. Wear earmuffs or earplugs when working on loud machines such as chain saws or lawn mowers.

Sturdy Work Boots: Wear steel-toe or heavy-duty boots to protect your feet from falling tools, engine parts, or debris.

Handling Fuel Safely

Fuel is one of the most dangerous aspects of small engine maintenance, and improper handling can lead to fires or explosions. Here are some tips for working safely with fuel.

Work in a Well-Ventilated Area: Always operate engines in a space that is well ventilated to avoid the buildup of fuel vapors, which can be hazardous.

Store Fuel Properly: Keep fuel in appropriate, clearly labeled containers that are designed for fuel storage. Always store fuel in a cool, dry place away from direct sunlight and heat sources.

Avoid Spills: Be careful when filling or draining fuel. Use funnels to avoid spills, and immediately clean up any spills using absorbent materials.

Fire Prevention

Fires can happen quickly if fuel or electrical components overheat. Here's how to reduce the risk.

Turn Off the Engine: Always make sure the engine is off and cool before performing maintenance, especially when working with electrical components or the fuel system.

Keep a Fire Extinguisher Nearby: Have a fire extinguisher rated for Class B (flammable liquids) and Class C (electrical) fires in your workspace.

Be Cautious With Hot Components: Parts of the engine, such as the exhaust system and spark plug, can become very hot during operation. Wait for these components to cool down before handling them.

By using the right tools, setting up an organized and safe workspace, and following essential safety practices, you can effectively maintain and repair small engines while minimizing the risk of injury. Proper tool usage and organization not only increases the efficiency of your repairs but also helps you avoid costly mistakes. Most importantly, always remember that safety comes first. By wearing appropriate protective gear, handling fuel properly, and preventing fire hazards, you can ensure a safe and productive repair experience every time.

CHAPTER 4

RESOURCES AND FURTHER LEARNING

As you advance in your small engine repair skills, it's critical to continue learning and evolving. The small engine repair industry is continually changing, with new technologies, tools, and techniques being developed all the time. Whether you're looking for reliable tools, informative books, or online resources, this chapter will provide you with a list of resources to help you expand your knowledge and improve your skills. Additionally, we will discuss certification programs and industry organizations that can enhance your credibility and keep you competitive. Finally, we will offer tips for staying up to date with new technologies to ensure you're always ready for the latest challenges

RECOMMENDED TOOLS, BOOKS, AND ONLINE RESOURCES

Recommended Tools for Small Engine Repair

Investing in the right tools is essential for running a successful small engine repair business or enhancing your skill set. Here's a list of tools that every small engine technician should have.

Basic Hand Tools: A well-stocked toolbox is the foundation of any small engine repair shop, including screwdrivers, pliers, adjustable wrenches, needle-nose pliers, and socket sets.

Recommendation: There are many brands of tools that are known for producing high-quality hand tools that will last for years.

Diagnostic Tools: Tools for diagnosing issues are critical for accurate repairs. A digital multimeter, compression tester, and spark plug tester will help you assess electrical and engine health.

Recommendation: A good multimeter is essential for checking circuits, batteries, and components.

Fuel System Tools: These tools help clean and repair carburetors, fuel lines, and injectors. Carburetor cleaning kits, gaskets, and fuel line repair kits will be necessary for tackling fuel system issues.

Recommendation: Look for brands that specialize in small engine parts and repair kits.

Specialty Tools: Depending on the type of small engine you work on, you may need specialized tools, such as a valve spring compressor, flywheel puller, or crankshaft tool, for rebuilding engines.

Recommendation: Several small engine manufacturers offer specific tools for their products, so it's worth investing in tools designed for specific brands, such as Honda, Briggs & Stratton, or Yamaha.

Workshop Equipment: Larger tools and equipment, such as a parts washer, air compressor, workbench, and engine hoist, will make your shop more efficient and organized.

Recommendation: A good-quality workbench and a durable air compressor are must-haves for handling bigger repair jobs.

RECOMMENDED BOOKS FOR LEARNING SMALL ENGINE REPAIR

When learning about small engine repair, one of the best ways to gain in-depth knowledge is to use factory service manuals. These manuals are specifically designed by manufacturers for technicians, making them an invaluable resource for both beginners and experienced mechanics. Factory service manuals provide detailed, accurate information about a specific engine model, including maintenance schedules, troubleshooting guides, repair procedures, parts diagrams, and torque specifications. Below are some of the most-recommended factory service manuals to aid in learning small engine repair.

Honda Small Engine Service Manual

Honda is renowned for producing durable, high-quality small engines that are commonly found in lawn mowers, generators, and various power equipment. Honda's factory service manuals are comprehensive and highly regarded in the industry. The manuals provide detailed explanations of engine components, diagnostic procedures, and repair techniques. They cover a range of models, including GX160, GX200, GX270, and GX390 engines, which are commonly used in small engines. For someone learning small engine repair, the Honda manual's step-by-step approach to engine disassembly, reassembly, and troubleshooting makes it an excellent resource.

Briggs & Stratton Service Manual

Briggs & Stratton is one of the most popular brands for small engines that are commonly found in lawn mowers, pressure washers, and other outdoor equipment. Its factory service manuals are invaluable for understanding how to maintain and repair its engines. These manuals contain detailed diagrams, procedures, and troubleshooting tips for engines such as the 5 HP to 20 HP models. The manuals also include information about carburetors, ignition systems, valve adjustments, and other key components. With clear instructions and easy-to-follow illustrations, Briggs & Stratton service manuals are ideal for learning engine repair, especially for those working on lawn and garden equipment.

Kohler Small Engine Service Manual

Kohler small engines are widely used in commercial and residential lawn mowers, generators, and other power equipment. Kohler factory service manuals provide extensive technical details about a wide range of small engines, from older models to the latest ones. The manuals cover the complete engine breakdown, offering a deeper understanding of how each component works together. Kohler service manuals also provide troubleshooting guidelines, maintenance routines, and step-by-step instructions for repairing or replacing various engine parts, making them an essential tool for anyone looking to learn engine repair.

Tecumseh Service Manual

Tecumseh is another key manufacturer of small engines, and its service manuals are widely used in the repair and maintenance of outdoor equipment engines. Tecumseh's factory service manuals include clear information about engine maintenance, component repair, and troubleshooting. These manuals cover a variety of models, including those used in snowblowers, lawn mowers, and generators. The Tecumseh manual offers instructions on disassembling and reassembling the engine, as well as valuable diagnostic steps to address common engine problems. Tecumseh's diagrams and parts breakdowns make it easier for learners to understand how the engine functions and how to perform maintenance effectively.

Kawasaki Small Engine Service Manual

Kawasaki offers reliable small engines that are often found in generators, pumps, and other equipment. Its service manuals are detailed and helpful, providing in-depth information about each engine's design and troubleshooting steps. Kawasaki manuals include clear procedures for tasks such as valve adjustments, carburetor cleaning, and overall engine inspection. For learners, these manuals also provide exploded views of engine assemblies, making it easier to visualize the repair process. Kawasaki factory service manuals are a great resource for anyone wanting to specialize in repair for their engines.

Factory service manuals are essential tools for anyone learning small engine repair. They provide not only the correct procedures for engine maintenance and repair but also offer diagnostic insights and detailed diagrams to help users understand engine functionality. Manuals from brands such as Honda, Briggs & Stratton, Kohler, Tecumseh, and Kawasaki are all highly recommended because they are specific to each manufacturer's engine models and provide step-by-step guidance for effective repair and maintenance. Using these manuals will enhance your repair skills and provide a deeper understanding of how small engines work, helping you become proficient in small engine repair.

ONLINE RESOURCES FOR SMALL ENGINE REPAIR

The internet is a treasure trove of information for small engine repair, offering forums, videos, tutorials, and more. Below are some of the top online resources for small engine repair.

YouTube

YouTube is an excellent place to watch step-by-step repair videos and find valuable repair tutorials and maintenance tips. Watching real-time repairs will give you practical insights into troubleshooting and fixing issues.

Small Engine Repair Forums

Online forums are fantastic for asking questions, sharing experiences, and getting advice from experienced professionals. These forums often feature discussions about the latest small engine technologies and repairs, as well as solutions to common issues.

Manufacturer Websites

Manufacturer websites such as Briggs & Stratton, Honda Engines, Kohler Engines, Tecumseh, and Kawasaki provide a wealth of resources, including manuals, troubleshooting guides, and parts diagrams. These websites are often the best place to find the most accurate and up-to-date repair information for specific engine models.

Online Courses and Tutorials

Websites such as Udemy, Skillshare, and Coursera offer courses on small engine repair. These platforms often provide in-depth lessons on diagnostics, maintenance, and repair, ranging from beginner to advanced levels.

Repair Websites and Apps

Websites offer comprehensive guides for fixing a variety of appliances and engines. These sites also provide troubleshooting tips, replacement parts, and detailed repair instructions.

Certification Programs and Industry Organizations

For those serious about pursuing a career in small engine repair, obtaining certifications and joining industry organizations can add credibility to your business and improve your skill set.

CERTIFICATION PROGRAMS

Becoming certified as a small engine technician can open new career opportunities, increase customer trust, and set you apart from other mechanics. Here are some of the top certification programs available.

Small Engine Repair Certification

Automotive Service Excellence (ASE) offers certification programs for small engine technicians. ASE certification is widely recognized in the industry and demonstrates that you have the technical knowledge and skills required for professional engine repair.

Briggs & Stratton Certified Technician Program

Briggs & Stratton offers a certification program that focuses specifically on its engines. This program is a valuable certification if you plan to work on or specialize in Briggs & Stratton engines, which are used in a wide range of applications.

Honda Small Engine Certification

Honda offers a certification program for technicians working with its engines. The program includes online and hands-on training, covering topics such as diagnostics, maintenance, and repair specific to Honda small engines.

Kawasaki Small Engine Certification

Kawasaki offers a certification program for technicians working with its engines. The program includes online and hands-on training, covering topics such as diagnostics, maintenance, and repair specific to Kawasaki small engines.

Certified Master Small Engine Technician (CMSET)

The Outdoor Power Equipment and Engine Service (OPEES) offers a Master Technician certification, which is a higher level of expertise recognized in the industry. This program covers all aspects of small engine repair, including troubleshooting, rebuilding, and repair of various engine types.

INDUSTRY ORGANIZATIONS

Joining industry organizations is another great way to stay updated and build connections in the small engine repair community. Here are some of the most prominent organizations.

Outdoor Power Equipment Institute (OPEI)

The Outdoor Power Equipment Institute (OPEI) is the leading trade association for the outdoor power equipment industry. The OPEI offers various resources for small engine professionals, including educational programs, conferences, and networking opportunities.

Engine Builders Association (EBA)

The Engine Builders Association (EBA) is an industry association dedicated to engine builders, repairers, and service providers. It offers workshops, webinars, and networking opportunities for professionals in the small engine repair field.

National Institute for Automotive Service Excellence (ASE)

In addition to their certifications, the National Institute for Automotive Service Excellence (ASE) offers valuable resources for mechanics, including newsletters, technical resources, and events where you can meet other industry professionals and learn about new technologies.

American Society of Mechanical Engineers (ASME)

For those interested in the engineering aspect of small engines, the American Society of Mechanical Engineers (ASME) offers professional development programs, certifications, and networking opportunities.

TIPS FOR STAYING UPDATED WITH NEW TECHNOLOGIES

The small engine repair industry, like others, is constantly evolving. New technologies and innovations in engine design, fuel efficiency, and diagnostics can have a significant impact on how you approach repairs. Here are some tips for staying updated.

Attend Workshops and Conferences

Attend industry conferences and workshops, such as those hosted by OPEI or ASE. These events are a great way to stay informed about the latest advancements in small engine technology, repair techniques, and regulations.

Subscribe to Industry Magazines and Journals

Subscribe to trade magazines such as *Outdoor Power Equipment Magazine* or *Engine Builder* for the latest news and trends in the small engine industry.

Follow Industry Leaders on Social Media

Small engine experts, manufacturers, and repair shops post valuable tips, troubleshooting guides, and product reviews on social media platforms. Following these accounts can help you stay informed in real-time.

CONTINUOUS LEARNING

Never stop learning. Make time for online courses, webinars, and tutorials to expand your knowledge and keep your skills sharp. Taking part in certification programs and seeking new training opportunities will help ensure you're always ready for the latest technologies.

As the small engine repair field continues to grow, staying informed about new tools, technologies, and techniques will help you stay ahead of the curve. By utilizing the right resources, pursuing certification programs, joining industry organizations, and keeping your skills up to date, you can ensure a long and successful career in the small engine repair industry. Whether you're an aspiring technician or a seasoned professional, the resources and strategies outlined in this chapter will support your journey toward mastery and success in the world of small engine repair.

"A Basic Guide to Small Engine Repair" is more than just a manual, it's a roadmap to gaining the confidence, skills, and knowledge you need to effectively maintain and repair small engines. Throughout the book, we've provided a comprehensive guide that covers everything from understanding the basic components of small engines to advanced troubleshooting, repair techniques, and even rebuilding engines from the ground up.

By mastering the step-by-step processes outlined in the chapters, you can develop the expertise needed to handle the wide range of small engine tasks you may encounter. From routine maintenance tasks, such as changing oil and replacing spark plugs, to diagnosing and fixing more complex issues, such as electrical faults or fuel system malfunctions, this guide ensures that you are well equipped to tackle it all.

One of the key takeaways from this book is that small engine repair does not have to be intimidating. By breaking down the repair process into manageable steps, even the most intricate repairs become more accessible. Understanding the fundamental components of the engine, as well as the role they play in its operation, is essential to identifying the root cause of common problems. Armed with the right knowledge, tools, and troubleshooting strategies, you can confidently approach any issue with your engine and work toward a solution.

Practical Applications and Real-World Skills

The practical skills you gain through this guide are not just theoretical, they are immediately applicable to everyday tasks. Whether you're fixing a chain saw to clear brush, repairing a generator to provide backup power, or servicing a lawn mower to keep your yard pristine, the techniques you will learn here can save you time, money, and frustration. Rather than relying on expensive repair services, you will have the ability to diagnose and solve problems on your own, keeping your equipment in top condition for longer.

Additionally, the book provides valuable insights into common small engine issues that so many people may not even realize are preventable. Regular maintenance is often the key to avoiding more costly repairs down the line. By doing routine tasks such as oil changes, air filter replacements, and spark plug checks as part of your regular routine, you will not only improve the performance of your engine, but you will also extend its lifespan.

GOING BEYOND THE BASICS

For those eager to expand their small engine repair expertise, this book goes beyond the basics by introducing advanced topics that delve deeper into the world of engine mechanics. It covers complex tasks such as rebuilding engines from the ground up, repairing specialty engines, and understanding the use of advanced diagnostic tools like multimeters, compression testers, and timing lights. These tools are essential for diagnosing issues with modern small engines, where electronic components and intricate systems are becoming increasingly common. With the guidance provided in this book, you will gain a clear understanding of how to use these tools to accurately assess engine performance, identify faults, and make precise repairs.

The inclusion of these advanced techniques opens up a wealth of new opportunities for mechanics and enthusiasts who want to work on larger, more complicated engine systems. Rebuilding an engine from scratch, for instance, requires a deep understanding of engine components, tolerances, and assembly processes. This book provides detailed instructions and tips for those looking to take on such projects with confidence, ensuring that every step is approached with accuracy and professionalism. Additionally, for those interested in modifying engines for improved performance, the book covers how to make modifications to various types of small engines, including chain saws, lawn mowers, and go-karts. Whether you're aiming for more power, efficiency, or specific customizations, the knowledge gained from these sections will allow you to tackle engine modifications with the skills needed to achieve the best results.

PART 2

MAINTENANCE AND TROUBLESHOOTING

CHAPTER 5

MAINTENANCE BASICS

Routine maintenance is essential to ensure that small engines continue to perform at their best, providing reliable power for years. Regular care can help prevent breakdowns, improve efficiency, and extend the lifespan of your equipment. Neglecting maintenance can lead to poor performance, costly repairs, and, ultimately, the need to replace the engine altogether.

This chapter covers the fundamentals of small engine maintenance, including the importance of routine upkeep and a step-by-step guide for performing some of the most common and critical maintenance tasks: changing oil, replacing air filters, checking and replacing spark plugs, and cleaning and adjusting carburetors.

Importance of Routine Maintenance: Maintaining small engines isn't just about keeping things running smoothly today, it's about preventing costly problems in the future. Routine maintenance helps.

Improve Engine Performance: Regularly maintaining components such as the air filter, spark plug, and carburetor ensures that the engine is getting the optimal fuel and air mixture it needs to run efficiently.

Increase Longevity: Just like any other machine, small engines require care to ensure they last. Routine checks and timely replacements of key parts can greatly extend the operational life of your engine.

Save Money: Proactively addressing minor issues before they become major ones can save you money on expensive repairs or replacements. Small engine repairs are typically more affordable when caught early, and maintenance can prevent you from needing a new engine prematurely.

Enhance Safety: A well-maintained engine is safer to operate. Components such as the fuel system, spark plug, and exhaust need to be in proper working order to avoid potential hazards such as fires, fuel leaks, or engine failure during use.

When performing maintenance on a small engine, it is important to use safety glasses to protect your eyes, gloves to protect your hands, and boots to protect your feet. When an engine is operating, earplugs, noise-canceling earbuds, or headphones are important to protect your hearing. According to the Occupational Safety and Health Administration (OSHA), loud noises produce decibels (dB), which are calculated using a time-weighted average over eight hours. Even if the dB is within OSHA tolerance, over time, your hearing may decline (2023).

When working on a small engine, it is important to remove the spark plug boot and wire that is connected to the spark plug. This practice will ensure that the engine will not accidentally start when performing maintenance. Even the simplest tasks require the spark plug boot to be disconnected from the spark plug. This step is a form of mechanical lockout when dealing with hazardous energy.

Steps Necessary To Change a Pre-Filter

1. Locate the air filter assembly. The air filter is typically located on the side or top of the engine. You may need to remove a plastic cover or shield to access it. The pre-filter will be found covering the primary air filter, which is usually a paper element. Check your user manual to locate the air filter on your specific engine model.

2. Remove the air filter cover. If the air filter is housed under a cover, use a screwdriver to remove the screws or fasteners securing it. Some covers are held in place by clips or tabs, while others may have screws or bolts. Carefully take off the cover to reveal the air filter.

3. Remove the old pre-filter. Once the air filter cover is off, the pre-filter should be visible around the air filter element. It may be a foam material or a mesh filter. Remove the old pre-filter from around the air filter element. It might simply pull off or slide off, depending on how it's installed.

4. Inspect the primary air filter. While you have the air filter assembly open, it's a good time to check the condition of the primary air filter. If it looks dirty, damaged, or clogged, it may need to be replaced or cleaned (depending on the type of filter). If the primary air filter is reusable, tap it gently to remove excess dirt or use compressed air to blow the debris out. If it's paper-based and too dirty, replace it with a new one.

5. Install the new pre-filter. Take the new pre-filter and carefully place it over the primary air filter, ensuring it fits snugly around the edges. If the pre-filter is foam, make sure it covers the entire filter element. If it's a mesh or fabric pre-filter, position it properly to allow air to flow while filtering out debris.

6. Reassemble the air filter cover. Once the new pre-filter is in place, replace the air filter cover. Ensure it aligns properly with the mounting holes and that it sits securely. Tighten any screws or fasteners to hold the cover in place.

7. Complete a final check. Double-check that everything is secured properly and ensure that the pre-filter is correctly installed. Start the engine to make sure everything is functioning smoothly. The engine should start without difficulty, and there should be no unusual air intake noises.

Steps Necessary To Change the Air Filter

1. Find the air filter cover.

2. Using the necessary hand tools (e.g., a screwdriver, socket wrench, etc.), unscrew the necessary fastener. Sometimes, the air filter cover is fastened with a wing nut that is hand tight.

3. Remove the old air filter and put it in the trash.

4. Clean the surface with contact cleaner where the old air filter was sitting.

5. Place the new air filter.

6. Replace the air filter cover.

Steps Necessary To Change Engine Oil

1. Place an oil pan under the engine where the oil will be released.

2. Using a socket wrench or open-end wrench, find the bolt or cap that is screwed into the engine block under the engine on a vertical shaft engine. Take out the bolt that is connected to the side of the engine on a horizontal shaft engine.

3. Drain the oil from the engine. If the engine is equipped with an oil filter, now is the time to unscrew that too. You may need to use an oil filter wrench. When the filter is off, make sure the rubber ring is not stuck on the engine block.

4. Replace the oil drain bolt or cap. Before replacing the oil filter (if the engine is equipped), fill the new oil filter with fresh oil. Additionally, lubricate the rubber O-ring on the filter too with

oil. This practice will help with ensuring that there is a good seal on the oil filter to the engine block.

5. Fill the engine with the recommended type and amount of oil. Refer to the engine's repair manual.

6. Clean any oily surfaces with contact cleaner.

7. Dispose of the used oil at a recycling facility that takes used oil or ask an automotive auto parts store to see if they take oil.

Steps Necessary To Gap a Spark Plug

1. Using a spark plug gapping tool, find the correct gap and use it on the spark plug.

Steps Necessary To Change a Spark Plug

1. Remove the spark plug boot.

2. Using a socket wrench, remove the old spark plug.

3. Purchase a new, correct spark plug.

4. Using a spark plug gapping tool, gap the spark plug using the engine's specific maintenance manual.

5. Replace the spark plug. If the spark plug is difficult to replace because of the spark plug hole being too far into the engine block, then a piece of spare fuel line can be handy to use. The end of the fuel line should be inserted into the end of the spark plug. Then, with a twisting motion, the spark plug will screw into the spark plug hole. Be sure that you don't cross-thread the spark plug.

Steps Necessary To Check Engine Compression

Note: A compression check should not be the sole indicator to see if the engine has enough compression. It is also important to perform a leak-down test of the cylinder. However, to complete a compression test, follow the directions below.

1. Remove the spark plug.

2. Find the correct engine compression adaptor and screw the adaptor into the spark plug hole.

3. Connect the compression gauge to the adapter and turn the engine over several times. Refer to the engine's maintenance manual for specific readings.

Steps Necessary To Sharpen a Lawn Mower Blade

1. Place a piece of wood (2x4 or 4x4) under the mower deck against the blade and mower deck to ensure the blade cannot move.

2. Using a socket and a breaker bar or impact wrench, loosen the bolt for the mower blade. Pay attention to how the blade, washers, and spacers come apart.

3. Clean old grass and debris from the mower blade.

4. Using a blade balancer, find which side of the blade is unbalanced and mark that side with a marker.

5. Using a file or grinding wheel, start to sharpen the blade to the contour of the blade itself. The blade does not need to be as sharp as a knife. It is important to make sure the consistency of

blade sharpening is equal on both sides of the blade and that the blade is balanced. A balanced blade will reduce unnecessary vibration to the engine and engine parts.

6. Reinstall the blade.

7. Reconnect the spark plug boot to the spark plug.

Steps Necessary To Clean a Carburetor

Note: Since carburetors come in all different shapes and sizes, the directions below are generalized.

1. Clamp off the fuel line, if necessary, and disconnect the fuel line from the carburetor.

2. Disconnect the carburetor from the engine. It is important to take pictures of the carburetor while it's connected to the engine.

3. Take the carburetor apart, paying attention to all of the parts that are disassembled. Again, take pictures.

4. Clean the carburetor with carburetor cleaner or soak the carburetor in parts cleaner.

5. After the parts are cleaned, reassemble the carburetor.

6. Reinstall the carburetor to the engine using new gaskets, if needed.

7. Reinstall the fuel line.

Steps Necessary To Adjust a Governor

Note: There are two types of governors: mechanical and air vane. A mechanical governor works using the camshaft, while an air vane governor works from the air that the flywheel expels. A governor should not need to be adjusted unless the governor's arm is tampered with. It is important to review the engine's maintenance manual to seek more information on governor adjustment.

Steps Necessary To Fix a Pull Rope Cord

1. The recoil starter is the component with the pull cord that you grab to start an engine. It is usually located on top of the engine or on the side. Depending on your model, it might be housed under a plastic cover, which you'll need to remove.

2. Remove the recoil starter assembly. Use a screwdriver or socket wrench to remove any screws or bolts securing the recoil starter assembly to the engine. Once the screws are removed, carefully take off the recoil starter cover. The pull cord mechanism is housed inside.

3. Inspect the pull cord mechanism. Check if the pull cord is broken or if the spring inside the recoil starter has come loose or is damaged. If the cord is simply worn or frayed, you can just replace it. If the spring is damaged, the spring mechanism might need to be replaced as well.

4. Remove the old cord (if applicable). If the cord is completely broken or you're replacing it, cut or untie the remaining pieces of the cord from the spool. Unwind the cord from the spool carefully, taking note of how it is wound and how the spring operates. The cord should be wound in a way that the spring is under tension.

5. Rewind the recoil spring (if necessary). If the spring is intact and you're just replacing the cord, you may need to rewind the recoil spring before installing the new cord. To tension the spring, with the recoil assembly exposed, rotate the pulley in the direction that will wind the spring. This movement will generate tension when you pull the cord. Be cautious of the spring tension because it can snap back quickly. If you're not sure, you can either use a spring rewind tool or gently apply force to get the right tension.

6. Cut and install the new pull cord. Measure the length by cutting a new pull cord to the correct length. The cord should be long enough to comfortably pull but not too long that it drags or gets tangled. To tie the knot, feed one end of the cord through the hole in the recoil starter pulley. Tie a secure knot at the end to prevent the cord from slipping through. To wind the cord, begin winding the cord around the pulley in the direction that it was originally wound (usually clockwise but check the old cord for reference). Keep the cord taut and avoid twisting it.

7. Reinstall the recoil starter assembly. Once the new cord is in place and the spring is properly tensioned, reattach the recoil

starter cover to the engine. Align the recoil starter assembly with the mounting holes and use screws or bolts to secure it back in place.

8. Test the pull cord. Before reattaching the spark plug, pull the rope to ensure it retracts smoothly and is under appropriate tension. Make sure the cord is retracting fully and that there is no resistance or jamming. If everything looks good, you can reattach the spark plug.

9. Complete a final check. Give the cord a few test pulls to make sure it operates properly. Ensure the lawn mower starts smoothly and the cord retracts without issue.

Steps Necessary To Replace a Fuel Filter

1. Clamp off the fuel line with special plastic clamps on each side of the fuel filter.

2. Disconnect the fuel filter.

3. Replace the fuel filter. Pay attention to the arrows. The arrows usually indicate the flow of the fuel from the fuel tank to the carburetor.

Two-Cycle: With needle-nose pliers or a small hook, scoop the fuel line and fuel filter from the fuel tank. Once the fuel filter is out of the tank, size it with a new fuel filter and replace it.

Steps Necessary To Replace a Fuel Line

1. Find the fuel line connected between the carburetor and the fuel tank.
2. Replace the necessary fuel line with the same exact diameter and the old fuel line.

Two-Cycle: With needle-nose pliers or a small hook, scoop the fuel line and fuel filter from the fuel tank. Once the fuel line is out of the tank, size it with a new fuel line and replace it. When replacing the fuel line, cut the end on the new fuel line at a sharp 45-degree angle. Place the 45-degree angle in the fuel tank. Once you can see the fuel line, grab it with a pair of needle-nose pillars. This step will save you time and hassle. Another trick is to spray a small amount of silicone spray on the fuel line to lubricate it to help it slide through the fuel tank.

Steps Necessary To Replace a Head Gasket

1. Remove the necessary pieces and parts that interfere with the head gasket.

2. Remove the bolts on the cylinder head. Make sure the bolts are kept in the same order on the cylinder head as they were loosened.

3. Rotate the engine so the piston is at the top of the cylinder. This position is called "Top Dead Center" (TDC).

4. Using a metal scraping tool, such as a putty knife, scrape off the head gasket, trying to keep it in one piece and being

careful not to damage the piston. Once the gasket is off, clean the head and top of the piston with a scouring pad.

5. Clean the inside gasket surface of the cylinder head with a scouring pad. You can also place the cylinder head in a sand blaster. Be very careful that the sand is not too abrasive.

6. If a sand blaster is used, the cylinder head will need to be cleaned with hot soap and water before installing it back to the cylinder block.

7. Install the head gasket and put the cylinder head back on the block. Torque the bolts to the recommended torque specifications per the specific engine's maintenance manual. Then, torque the bolts in a crisscross pattern, making sure that each side of the cylinder head has the same amount of tension on it.

8. Replace all other parts that were taken off to remove the cylinder head. Make sure that all bolts that were taken off are replaced at the proper torque.

Steps Necessary To Adjust Engine Idle

1. Find the idle adjustment screw if the carburetor has one.

2. Turn the screw clockwise very slightly until the engine is at its desired idle speed.

3. Wait a few seconds to ensure that the RPMs are at the desired speed.

Steps Necessary To Change a Battery

1. Make sure the lawn tractor is on a flat surface and the ignition is turned off with the key out of the key switch.

2. Disconnect the negative battery terminal (black terminal).

3. Disconnect the positive battery terminal (red terminal).

4. Remove the old battery.

5. Clean the battery compartment.

6. Use a battery terminal cleaner or a wire brush to remove corrosion from the terminals.

7. Place the new battery into the compartment, ensuring the positive and negative terminals are correctly positioned.

8. Reconnect the positive battery terminal.

9. Start by reconnecting the positive (+) cable to the positive terminal on the new battery. Tighten the bolt securely but avoid over-tightening.

10. Reconnect the negative battery terminal. Tighten the bolt securely.

11. Test the new battery.

12. Turn the ignition key to the "on" position to test if the tractor starts. If you have a multimeter, you can check the battery voltage to make sure it's between 12.6V and 12.8V when fully charged.

13. Dispose of the old battery. Properly dispose of the old battery at a recycling center or a location that accepts used batteries. Several auto parts stores also offer battery recycling.

Steps Necessary To Fill a Battery

1. Make sure the tractor's engine is off and remove the key from the ignition to prevent accidental starts.

2. Inspect the battery. Look at the battery to see if the electrolyte (the liquid inside) level is low. The battery should have a series of vent caps or plugs that cover the cells. Remove the battery caps (if applicable).

3. Check the electrolyte level. Inside the battery, you'll see the electrolyte liquid, which is a mix of sulfuric acid and water. The level should cover the internal lead plates. If the fluid level is low and the plates are exposed, you will need to add distilled water.

4. Add distilled water. Using a clean funnel, add distilled water slowly to each cell. Only use distilled water – tap water contains minerals that can damage the battery. Add water until the electrolyte covers the lead plates inside the battery, but don't overfill. There should be a visible gap between the water level and the top of the cell.

5. Replace the caps. After filling, carefully replace the battery caps. Ensure they are securely tightened to prevent any leakage or contamination.

6. Check the battery terminals. Inspect the battery terminals for corrosion. If corrosion is present, clean the terminals with a mixture of baking soda and water, then dry thoroughly. Apply a thin layer of petroleum jelly or terminal grease to prevent further corrosion.

7. Charge the battery (if necessary). If the battery was low on water, it may also be low on charge. Use a battery charger to charge the battery fully, following the manufacturer's instructions for your tractor's battery.

8. Test the tractor. Reinstall the battery into its compartment if you remove it. Turn the tractor's ignition key to the "on" position and start the engine. Ensure the tractor starts smoothly and that the electrical components are functioning properly.

9. Check for leaks. After use, check the battery area for any signs of leaking or acid buildup. Clean the area if necessary.

Steps Necessary To Sharpen a Saw Chain

1. Inspect the chain. Check the chain for damage or wear. If the chain is excessively worn or damaged (e.g., teeth are chipped), it might need to be replaced. Make sure the chain is properly tensioned before you start sharpening. If the chain is too loose, it can make the sharpening process difficult and uneven.

2. Secure the chain saw. Place the chain saw securely in a vice or on a sturdy work surface. Ensure the bar and chain are positioned so you can easily access the cutting teeth. If using a vice, clamp the chain saw by the bar, not the chain, to avoid damage.

3. Clean the chain. Clean the chain using a rag to remove any dirt, sap, or

debris. This step ensures you can see the teeth clearly and have a clean workspace.

4. Identify the cutting teeth. Look at the teeth on the chain. Each tooth has a sharp edge that is used to cut wood. These teeth are alternately angled to the left and right. Work on one side of the chain at a time (either left or right), then rotate the chain to sharpen the opposite side.

5. Position the round file. Select a round file that matches the diameter of the chain's teeth. Chain saw chains come in different sizes, so be sure to use the correct file size (check the manufacturer's specs for your saw). Place the file into the cutting tooth at the proper angle (usually around 30-35 degrees for most chains). File the tooth by pushing the file across the tooth in a forward motion, maintaining a consistent angle. Apply firm, even pressure as you file. Each stroke should be smooth and consistent. Sharpen each tooth 2-3 times, or as needed. The goal is to restore a sharp edge to each tooth, so the cutting surface should feel smooth to the touch after filing.

6. Rotate the chain. After sharpening all the teeth on one side of the chain, rotate the chain to the next section. Switch to the other side of the chain and repeat the filing process, following the same steps. Make sure you file each tooth on the opposite side with the same number of strokes to keep the chain balanced.

7. Check the depth gauges. After sharpening the teeth, check the depth gauges (the small protrusions in front of each tooth)

that control how deep the tooth cuts into the wood. If the depth gauges are too high, they will prevent the teeth from cutting efficiently. Use a flat file to lower the depth gauges if necessary. Typically, a depth gauge tool can be used to measure the proper height of these gauges.

8. Recheck the chain tension. After sharpening the chain, check the tension again. A properly tensioned chain is essential for smooth and effective cutting.

9. Test the chain. Once sharpening is complete, reinstall the spark plug wire and start the chain saw. Test the chain by making a few cuts in a piece of wood. The saw should cut smoothly and easily. If it doesn't, check the chain again for any missed areas or uneven sharpening.

CHAPTER 6

TROUBLESHOOTING COMMON PROBLEMS

Even with proper maintenance, small engines can experience problems from time to time. Understanding how to diagnose and troubleshoot these issues is essential for efficient repairs and maintaining engine performance. Whether it's an engine that won't start, stalls while running, or simply runs rough, knowing how to identify the symptoms and systematically diagnose the problem can save you both time and money.

This chapter will guide you through the common symptoms of small engine problems, provide a diagnostic checklist to help identify fuel-, spark-, and air-related issues, and teach you how to interpret signs of wear or damage to key engine components.

Identifying Symptoms of Common Small Engine Problems

Understanding the specific symptoms of engine issues is the first step toward effective troubleshooting. Some of the most common symptoms you'll encounter include the following.

1. Engine Won't Start

One of the most frustrating problems is when an engine refuses to start. This issue can be caused by various factors, including problems with the fuel system, ignition, or even compression. If your engine doesn't start, consider the following.

Fuel-Related Issues: Old, stale, or contaminated fuel can cause starting problems. Additionally, a clogged fuel filter or fuel line can prevent fuel from reaching the engine.

Spark Issues: A faulty spark plug or ignition system may fail to ignite the fuel-air mixture, making it impossible for the engine to start.

Battery Issues (in electric start engines): If the battery is dead or damaged, the engine won't turn over, even if the fuel and ignition systems are working properly.

2. Engine Stalls or Cuts Off

An engine that starts but then stalls or cuts off during operation can be frustrating. This situation can be caused by a variety of issues, including a lack of fuel, ignition failure, or air supply issues.

Fuel Delivery Problems: A clogged fuel filter or carburetor can cause fuel delivery to be inconsistent, leading to stalling.

Dirty Air Filter: A blocked or clogged air filter can restrict airflow, causing the engine to stall as it struggles to run.

Ignition Issues: A worn-out or faulty spark plug may cause intermittent ignition, leading to stalling.

3. Engine Runs Rough or Misfires

If the engine runs unevenly, sputters, or misfires, it can be caused by poor combustion. This problem may be due to issues with fuel, spark, or air.

Fuel Mixture Problems: A carburetor that is dirty or improperly adjusted may provide a lean (too much air) or rich (too much fuel) mixture, causing the engine to run rough.

Spark Plug Issues: A spark plug that is dirty, worn, or improperly gapped can cause weak or inconsistent sparks, leading to misfires or rough running.

Air Supply Problems: A clogged air filter or intake valve can starve the engine of air, causing poor combustion and rough running.

4. Poor Acceleration

If the engine hesitates or has poor acceleration, it may indicate an issue with the carburetor, fuel delivery, or ignition system.

Clogged Carburetor Jets: If the carburetor is clogged or dirty, it may not deliver enough fuel at higher speeds, causing poor acceleration.

Fuel Line Issues: A kinked or clogged fuel line can restrict fuel flow, leading to insufficient power during acceleration.

Ignition Timing Issues: If the spark timing is off, the engine may misfire under load, causing hesitation during acceleration.

Diagnostic Checklist for Fuel, Spark, and Air Issues

Once you have identified the symptoms, it's time to systematically troubleshoot the engine's key systems: fuel, spark, and air. These three systems must work together to ensure smooth engine operation.

Fuel System

A number of problems can occur within the fuel system that can prevent the engine from running efficiently. Here's how to diagnose fuel-related issues.

Check the Fuel

Fresh Fuel: Ensure that the fuel in the tank is fresh. Stale fuel can cause poor engine performance or failure to start. If the fuel is old (more than 45 days), drain the tank and replace it with fresh fuel.

Fuel Contamination: Inspect the fuel for water, debris, or contaminants. If contamination is present, drain the fuel system and clean it out.

Inspect the Fuel Lines

Clogged or Kinked Lines: Inspect the fuel lines for cracks, clogs, or kinks. If you find any damage, replace the fuel lines.

Fuel Filter: Check the fuel filter for dirt or blockages. A clogged fuel filter can restrict fuel flow and cause starting or running problems. Replace the filter if needed.

Carburetor Issues

Clogged Jets or Passages: Dirt or debris in the carburetor can clog the jets, preventing proper fuel flow. Remove and clean the carburetor, using carburetor cleaner to remove deposits and dirt. Be sure to follow the manufacturer's instructions for cleaning.

Fuel Mixture: If the carburetor is adjustable, ensure the air-fuel mixture is properly set. If the engine runs rich (too much fuel) or lean (too much air), adjust the carburetor as needed.

Spark System

A weak or absent spark can cause the engine to misfire or fail to start. Here's how to check the spark system.

Inspect the Spark Plug

Check for Fouling: Remove the spark plug and inspect it for carbon deposits, oil, or dirt. A dirty spark plug may cause a weak spark or no spark at all. Clean or replace the spark plug as needed.

Check the Gap: Use a spark plug gap tool to check the spark plug gap. If it is too wide or too narrow, it can cause poor performance. Adjust the gap to the manufacturer's specifications.

Spark Plug Condition: If the spark plug is damaged, cracked, or excessively worn, then replace it.

Test the Ignition Coil

Ignition Coil Issues: If the spark plug is in good condition but you're still having problems with the ignition, the ignition coil may be faulty. Test the coil with a multimeter to check its resistance. If it's outside the recommended range, it will need to be replaced.

Check the Spark

Spark Test: Use a spark tester to check for a strong spark when the engine is cranked. A weak or intermittent spark could indicate a problem with the ignition coil or wiring.

Air System

The air system ensures that the engine gets enough oxygen for combustion. A lack of proper airflow can cause the engine to run poorly or stall. Here's how to troubleshoot the air system.

Check the Air Filter

Inspect for Clogs: Remove the air filter and inspect it for dirt, debris, or blockages. A clogged air filter can restrict airflow and cause the engine to stall or run rough. Clean or replace the filter as necessary.

Check for Damage: Inspect the air filter for any tears or damage. A damaged filter won't properly filter out dirt and debris and may need replacement.

Inspect the Carburetor's Air Intake

Check for Blockages: The carburetor's air intake may be blocked by dirt or debris, restricting airflow. Clean the intake and carburetor housing to ensure proper airflow.

Check the Valve and Compression

Intake Valve: If the engine has an intake valve, ensure it is opening and closing properly. A stuck intake valve can prevent the engine from getting enough air.

Compression Test: If the engine still runs poorly despite a clean air filter and intake system, perform a compression test to check the engine's overall condition. Low compression could indicate internal engine damage, such as worn piston rings or a damaged valve.

How To Interpret Signs of Wear or Damage

As you troubleshoot, keep an eye out for common signs of wear or damage that can indicate deeper issues. Here are some key indicators to watch for during your inspection.

Worn Spark Plug

A worn spark plug may exhibit signs such as corrosion, heavy carbon buildup, or a gap that is too wide. If cleaning the spark plug doesn't resolve the issue, replace it with a new one.

Clogged Carburetor Jets

If the carburetor is clogged, the engine may run rich or lean, causing rough running or stalling. Cleaning the carburetor may restore normal performance, but severe clogging may require replacing the carburetor or specific parts like the jets.

Fuel Line Wear

Cracked, brittle, or clogged fuel lines can cause fuel delivery issues. If you notice visible cracks or fuel leakage, replace the fuel line immediately.

Low Compression

Low compression can be a sign of worn piston rings, damaged valves, or head gasket failure. Low compression can cause poor performance, difficulty starting, or stalling. Performing a compression test will help confirm the diagnosis, and fixing low compression often requires more extensive repairs, such as replacing the piston rings or valves.

Air Filter Damage

A damaged or excessively dirty air filter can cause poor airflow, leading to poor engine performance. If the filter is torn or clogged beyond repair, replace it.

Diagnosing and troubleshooting small engine problems requires a methodical approach, starting with identifying symptoms and moving through a series of checks for fuel, spark, and air issues. By following a diagnostic checklist and interpreting signs of wear or damage, you can identify and address common problems that affect engine performance.

CHAPTER 7

TROUBLESHOOTING FAQS

Troubleshooting small engine problems can be a daunting task, especially if you're not sure where to begin. Fortunately, most small engine issues can be diagnosed and fixed with a little patience, basic knowledge, and the right approach. In this chapter, we will address some of the most common questions and issues that small engine owners face. We will provide quick fixes, helpful tips, and flowcharts to guide you through the process. Additionally, we will explore when it's best to seek professional help to ensure that your engine gets the care it needs.

COMMON QUESTIONS AND QUICK FIXES

Small engines, whether in lawn mowers, chain saws, or generators, are known for their reliability, but like any mechanical system, they can develop problems over time. Here are some of the most frequently asked questions regarding small engine issues, along with some quick fixes.

1. Why Won't My Engine Start?

One of the most common and frustrating issues with small engines is when they refuse to start. The reasons for this problem can vary, but the following are the most likely causes.

A. Fuel Issues: Stale fuel may be the culprit. If fuel has been sitting in the engine for a long period (especially over the winter), it may have broken down and caused clogging in the carburetor.

1. Quick Fix: Drain the old fuel and replace it with fresh fuel. Consider adding a fuel stabilizer to prevent future breakdown. Make sure you drain the fuel out of the carburetor bowl too. This part is the round-shaped object at the bottom of the carburetor.

B. Spark Plug Problems: A dirty or faulty spark plug can prevent the engine from firing. Over time, spark plugs can become covered in carbon deposits, or their electrodes may wear out.

1. Quick Fix: Remove the spark plug, inspect it for wear or damage, and clean it with a wire brush. If it's heavily damaged, replace it with a new one.

C. Fuel Line Blockages: Dirt, debris, or old fuel can clog the fuel line, stopping fuel from reaching the carburetor.

1. Quick Fix: Inspect and clean the fuel lines and replace any cracked or worn hoses.

D. Air Filter Clogging: A clogged air filter restricts airflow, which can prevent the engine from starting or cause it to run poorly.

1. Quick Fix: Check the air filter and clean it or replace it if it's too dirty or damaged.
2. Why Does My Engine Run Rough or Stall?

If your engine starts but runs roughly or stalls frequently, the issue is often related to fuel, spark, or air systems.

A. Carburetor Problems: Dirt or old fuel can cause clogging in the carburetor, leading to poor fuel delivery or a lean fuel mixture.

1. Quick Fix: Clean the carburetor thoroughly and check the jets for blockages. If cleaning doesn't resolve the issue, the carburetor may need to be rebuilt or replaced.

B. Fuel System Leaks: A leak in the fuel system, such as a crack in the fuel line, can cause fuel to escape, leading to inconsistent engine performance.

1. Quick Fix: Inspect the fuel system for leaks. Replace any damaged fuel lines or seals.

C. Excessive Vibration: Excessive vibration can cause the engine to stall or run poorly, often because of loose or damaged components such as the flywheel or the engine mounting bolts.

1. Quick Fix: Tighten any loose bolts, nuts, or screws. Check for loose or worn engine mounts and replace them if necessary.

D. Incorrect Carburetor Settings: Incorrect fuel-air mixture settings can lead to rough running.

1. Quick Fix: Adjust the carburetor's idle and mixture screws according to the manufacturer's specifications.

3.Why Is My Engine Overheating?

Overheating can be a serious issue, leading to engine damage if it's not addressed quickly. Common causes include poor airflow, insufficient oil, or cooling system blockages.

A. Low Oil Levels: Insufficient oil can cause the engine to overheat and seize.

1. Quick Fix: Check the oil level and top it off, if necessary. If the oil is dirty, change it immediately.

B. Clogged Cooling Fins: Dirt, debris, or grass can build up around the engine's cooling fins, reducing airflow and causing the engine to overheat.

1. Quick Fix: Clean the cooling fins with a soft brush or compressed air to ensure proper airflow.

C. Blocked Air Intake: A blocked or clogged air filter can also contribute to engine overheating.

1. Quick Fix: Inspect and clean or replace the air filter.

D. Faulty Thermostat: On some engines (especially in outboard motors or generators), a malfunctioning thermostat can cause overheating by failing to regulate the engine's temperature.

1. Quick Fix: Replace the thermostat if it is faulty.

READER-FRIENDLY FLOWCHARTS FOR PROBLEM-SOLVING

To help you diagnose and fix engine problems quickly, here are some simple flowcharts that break down the most common issues into manageable steps. Follow these flowcharts for fast problem-solving.

Engine Won't Start

1. Is the fuel tank empty?

   Yes: Refill the fuel tank with clean fuel.

   No: Proceed to the next step.
2. Is the fuel line clogged or kinked?

   Yes: Clean or replace the fuel lines.

   No: Proceed to the next step.
3. Is the spark plug dirty or faulty?

   Yes: Clean or replace the spark plug.

   No: Proceed to the next step.
4. Is the ignition coil faulty?

   Yes: Test the ignition coil with a multimeter and replace, if needed.

   No: Proceed to the next step.
5. Are the choke/throttle settings correct?

   No: Adjust the choke/throttle settings.

   Yes: Proceed to the next step.

6. Is the fuel stale or old?

   Yes: Drain the tank and replace it with fresh fuel.

   No: Proceed to the solution.

   Solution

   The engine should now start. Ensure all repairs and adjustments are completed.

Engine Starts but Runs Rough

1. Is the air filter dirty or clogged?

   Yes: Clean or replace the air filter.

   No: Proceed to the next step.

2. Is the carburetor clogged?

   Yes: Clean the carburetor and ensure the jets are clear.

   No: Proceed to the next step.

3. Are there fuel mixture issues?

   Yes: Adjust the carburetor to the proper mixture settings.

   No: Proceed to the next step.

4. Is the spark plug in good condition?

   No: Replace the spark plug.

   Yes: Proceed to the solution.

   Solution

   The engine should now run smoothly. Ensure all repairs and adjustments are completed.

Engine Stalls or Dies After Starting

1. Are there fuel delivery issues?

   Yes: Check the fuel filter and fuel lines for blockages.

   No: Proceed to the next step.

2. Is there a weak spark?

   Yes: Test the ignition system and replace faulty components.

   No: Proceed to the next step.

3. Is the engine overheating?

   Yes: Clean the cooling fins/vents and check the cooling fan.

   No: Proceed to the next step.

4. Is the oil level low?

   Yes: Ensure there's adequate oil.

   No: Proceed to the solution.

   Solution

   The engine should now run properly. Ensure all repairs and adjustments are completed.

Engine Makes Unusual Noises

1. Are there loose parts?

   Yes: Tighten any loose parts or bolts.

   No: Proceed to the next step.

2. Is the oil level low?

   Yes: Check the oil level and top off or replace it.

   No: Proceed to the next step.

3. Are the bearings damaged?

   Yes: Inspect the worn bearings and replace them, as needed.

No: Proceed to the solution.

Solution

The engine should now be running smoothly. Ensure all repairs and adjustments are completed.

Engine Is Smoking

1. Is the oil overfilled?

   Yes: Drain the excess oil.

   No: Proceed to the next step.

2. Is the air filter clogged?

   Yes: Clean or replace the air filter.

   No: Proceed to the next step.

3. Is the engine burning oil?

   Yes: Inspect the piston rings or valve seals and replace them, if necessary.

   No: Proceed to the solution.

   Solution

   The engine should now run without smoke. Ensure all repairs and adjustments are completed.

Engine Is Hard To Start or Turns Over Slowly

1. Is the battery weak or dead?

   Yes: Charge or replace the battery.

   No: Proceed to the next step.

2. Are the battery terminals corroded?

   Yes: Clean or replace the corroded battery terminals and check the wiring connections.

   No: Proceed to the next step.

3. Is the starter motor dirty or faulty?

   Yes: Clean or replace the starter motor.

   No: Proceed to the solution.

   Solution

   The engine should now start correctly. Ensure all repairs and adjustments are completed.

Engine Is Leaking Fuel

1. Is the fuel line cracked or damaged?

   Yes: Replace any damaged fuel lines.

   No: Proceed to the next step.

2. Is the fuel shutoff valve faulty or leaking?

   Yes: Inspect and replace the fuel shutoff valve if it's leaking.

   No: Proceed to the next step.

3. Is the fuel cap loose or not sealing?

   Yes: Tighten or replace the fuel cap.

   No: Proceed to the solution.

   Solution

   The engine should no longer leak fuel. Ensure all repairs and adjustments are completed.

Engine Backfires

1. Is the ignition timing incorrect?

   Yes: Check and adjust the ignition timing.

   No: Proceed to the next step.

2. Is the carburetor clogged?

   Yes: Clean the carburetor and adjust the fuel mixture.

   No: Proceed to the next step.

3. Is the spark plug dirty or worn?

   Yes: Replace the spark plug.

   No: Proceed to the solution.

   Solution

   The engine should no longer backfire. Ensure all repairs and adjustments are completed.

WHEN TO SEEK PROFESSIONAL HELP

While small engine issues can be resolved with basic tools and troubleshooting, there are times when seeking professional help is necessary. Here are some situations in which you should call in an expert.

Complex Carburetor Problems

If your carburetor is heavily gummed up or clogged, or if you're unable to fix the problem by cleaning it, it may require a professional rebuild or replacement. A carburetor that's out of alignment or damaged may also need expert calibration to get it running correctly.

Compression Issues

If you suspect that your engine has low compression (for example, if it cranks but doesn't start or lacks power), diagnosing the issue can require special equipment such as a compression tester. Low compression may be due to worn piston rings, a blown gasket, or damage to the cylinder, all of which may require disassembling the engine. A professional mechanic can perform these repairs and ensure everything is reassembled properly.

Electrical System Malfunctions

If you're having trouble with the ignition system, such as a faulty ignition coil or issues with the wiring, this problem can be tricky to troubleshoot. Diagnosing and repairing electrical components such as the stator, kill switch, or wiring requires specialized knowledge and testing equipment.

Major Engine Damage

If your engine is showing signs of severe wear or damage, such as metal shavings in the oil, excessive smoke, or strange noises, then it may be time to seek a professional's opinion. Internal engine repairs, including fixing a cracked block or replacing major components such as the crankshaft, require advanced skills and can be expensive. A mechanic can assess whether rebuilding the engine is worthwhile or if replacement is the better option.

Lack of Time or Tools

If you don’t have the time, tools, or confidence to tackle engine repairs yourself, then hiring a professional is often the best course of action. They can quickly diagnose the issue, provide a detailed estimate, and perform repairs efficiently, saving you both time and stress.

Troubleshooting small engine problems can seem overwhelming at first, but with a systematic approach, most issues can be identified and fixed quickly. By using the provided flowcharts and troubleshooting steps, you can diagnose common problems, such as starting issues, rough running, and overheating. However, for more complex issues or when in doubt, don’t hesitate to seek professional assistance. With a little practice and the right resources, you’ll become more confident in maintaining and repairing your small engine, ensuring it runs smoothly for years to come.

CHAPTER 8

FUEL SYSTEM REPAIR

The fuel system is one of the most critical components of a small engine, responsible for delivering the right mixture of fuel and air to the engine's combustion chamber. A well-maintained fuel system ensures optimal engine performance, while a malfunctioning fuel system can cause a host of problems, including hard starting, stalling, or poor fuel efficiency.

In this chapter, we will delve into the basics of fuel systems, including gasoline engines, carburetors, and fuel injection systems. We will also guide you through the process of cleaning and rebuilding carburetors, diagnosing fuel line issues, and providing solutions to keep your small engine running smoothly.

UNDERSTANDING FUEL SYSTEMS

Small engines typically use either a carburetor or a fuel injection system to mix air and fuel in the correct ratio before delivering it to the combustion chamber. Both systems have the same fundamental goal: to provide the engine with a consistent supply of air-fuel mixture that is necessary for combustion.

However, the design and operation of the two systems are quite different.

Gasoline Engines and Fuel Types

Small engines generally run on gasoline, a combustible liquid fuel derived from crude oil. Gasoline engines rely on a precise mixture of air and gasoline to produce power. The gasoline in small engines is usually mixed with oil in two-stroke engines, while four-stroke engines separate the oil and fuel.

Unleaded Gasoline: Most small engines today are designed to run on regular unleaded gasoline (89-94 octane). Using fuel with too low an octane rating can result in knocking and poor engine performance, while higher-octane fuels typically offer little added benefit and can be more expensive.

Ethanol-Blended Gasoline: Small engines are designed to run on gasoline that contains up to 10% ethanol (E10). However, it is important to note that if an engine sits too long with ethanol in the carburetor, it can lead to fuel degradation and possible carburetor and fuel line problems, especially in engines that aren't used frequently. For engines that sit unused for long periods, it's often recommended to use non-ethanol fuel.

Pre-engineered Fuels: Ideally, small engines mostly benefit from fuels that are pre-engineered. This type of fuel is ethanol free.

Carburetors

A carburetor is the traditional fuel delivery system used in small engines. It is responsible for mixing the right amount of fuel with air before sending the mixture into the combustion chamber. The carburetor uses a system of jets, valves, and metering rods to control the flow of air and fuel.

Function of the Carburetor: The carburetor's primary job is to ensure the correct air-to-fuel ratio. The engine requires a rich mixture at startup, transitioning to a leaner mixture during idle and high-speed operation.

Common Carburetor Issues: Carburetors can develop problems because of dirt or debris blocking the jets, which can prevent proper fuel flow. In addition, the gaskets, seals, and internal components can degrade over time, causing fuel leakage or poor mixture control.

Fuel Injection Systems

Fuel injection is a more modern fuel delivery system that has gradually replaced carburetors for several manufacturers of newer small engines, including those used in lawn mowers, chain saws, and other small power equipment. Fuel injection systems use electronic sensors and injectors to precisely meter fuel into the engine's intake air stream.

Advantages of Fuel Injection: Fuel injection systems are more efficient and deliver more precise control over the fuel mixture, leading to improved fuel economy, lower emissions, and better overall engine performance. These systems typically have fewer maintenance issues than carburetors but can be more expensive to repair.

Common Fuel Injection Problems: Fuel injectors can become clogged or dirty, leading to uneven fuel delivery or a misfire. Fuel pump issues or a faulty electronic control module (ECM) can also lead to starting or running problems.

Clean the Carburetor

Clean the Float Bowl: Use a carburetor cleaner to spray inside the float bowl and all the internal parts. Allow the cleaner to soak for a few minutes to loosen any dirt or debris. For stubborn deposits, use a soft brush to scrub.

Clean the Jets: Remove the jets from the carburetor and soak them in carburetor cleaner. Use compressed air to blow out any dirt or debris.

Soak the Carburetor: If the carburetor is very dirty, you may want to soak it in a carburetor cleaner solution for 15-30 minutes. After soaking, scrub it again to remove any remaining deposits.

Inspect Components for Wear: Check the needle valve, gaskets, and seals for wear. If any parts are damaged, replace them with new ones.

Reassemble the Carburetor

Once the carburetor is clean, reassemble it by carefully reinstalling the jets, needle valve, float bowl, and other components. Replace any worn-out gaskets with new ones to prevent fuel leakage.

Install the Carburetor

Once the carburetor is reassembled, reinstall it onto the engine, ensuring that it is properly secured. Reconnect the fuel line and fuel filter and check for any fuel leaks before starting the engine.

Tips for Carburetor Maintenance

Use Fresh Fuel: Always use fresh fuel because old fuel can cause the carburetor to clog.

Regular Cleaning: If you notice performance issues such as rough idling or difficulty starting, clean the carburetor regularly.

Add Fuel Stabilizer: When storing small engines for an extended period, add a fuel stabilizer to prevent fuel from deteriorating.

Diagnosing and Fixing Fuel Line Issues

Fuel lines are critical in transporting gasoline from the fuel tank to the carburetor or fuel injectors. Over time, fuel lines can become cracked, brittle, clogged, or detached, leading to a variety of problems such as fuel leakage, poor engine performance, or engine failure.

Common Fuel Line Problems

Cracked or Leaking Fuel Lines

Fuel lines can crack or become brittle because of age, exposure to sunlight, or extreme temperatures. Cracks in the fuel line can cause fuel leaks, which are dangerous and can lead to fire hazards.

Clogged Fuel Lines

Dirt, debris, and old fuel can clog fuel lines over time, restricting fuel flow to the engine and causing performance issues.

Detached Fuel Lines

Loose or detached fuel lines can result in fuel leakage or an inadequate fuel supply to the engine, leading to starting problems or stalling.

STEPS TO DIAGNOSE AND FIX FUEL LINE ISSUES

Inspect the Fuel Lines

Visually inspect the entire length of the fuel lines for any cracks, leaks, or damage. If you notice fuel stains around the fuel line connections, it's likely that there's a leak. Fuel lines should also be flexible and free from kinks. If the fuel lines are cracked or damaged, replace them.

Clean the Fuel Lines

If the fuel lines appear dirty or clogged, disconnect them from the fuel tank and carburetor, then carefully flush them with carburetor cleaner or compressed air. Be sure to clear out any dirt or debris that might be blocking the lines.

Replace Fuel Lines

If the fuel lines are cracked or beyond repair, replace them with new ones. Ensure that you use the correct size and type of fuel line for your engine model. It's best to use fuel lines made of durable material that can withstand the harsh chemicals in gasoline.

Check the Fuel Filter

The fuel filter is an important component that prevents dirt and debris from entering the carburetor. If the fuel filter is clogged, it can restrict fuel flow and cause engine problems. Replace the fuel filter if necessary. When replacing the fuel filter, pay attention to the arrow on the filter itself. The arrow indicates the actual flow from the gas tank to the carburetor.

Reassembly and Test

After replacing or cleaning the fuel lines, reassemble the fuel system and test the engine to ensure it is receiving an adequate fuel supply. Check for any fuel leaks around the fuel line connections and the carburetor.

Fuel system repair is a critical skill for maintaining small engines. Whether you're cleaning and rebuilding a carburetor or diagnosing fuel line issues, a properly functioning fuel system ensures optimal engine performance and reliability. Regular maintenance, such as using fresh fuel, inspecting fuel lines, and cleaning the carburetor, will help you avoid common fuel-related problems and keep your small engine running smoothly.

CHAPTER 9

ELECTRICAL SYSTEM REPAIR

The electrical system in a small engine is essential for starting and operating the engine efficiently. At the heart of this system lies the ignition system, which includes components such as the spark plug, ignition coil, kill switch, and wiring. When one of these parts fails or malfunctions, the engine may not start, or it could experience poor performance, misfires, or irregular operation. This chapter will provide an overview of small engine ignition systems, guidance on diagnosing and replacing spark plugs and ignition coils, and an explanation on how to understand and test kill switches and wiring.

OVERVIEW OF IGNITION SYSTEMS

The ignition system is responsible for providing the spark needed to ignite the air-fuel mixture in the engine's combustion chamber. Without a functional ignition system, a small engine simply cannot start or operate.

There are generally two types of ignition systems used in small engines: magneto ignition and battery-powered ignition systems.

Magneto Ignition System

A magneto ignition system is common in small engines, especially older models or those that do not require a battery. In a magneto system, the ignition coil is powered by the flywheel, which contains magnets that generate an electrical current as the flywheel rotates. This current is then sent to the ignition coil, which steps up the voltage and sends it to the spark plug, creating the spark that ignites the fuel-air mixture. The advantage of a magneto system is that it doesn't rely on an external power source (like a battery), which makes it ideal for smaller, simpler engines.

Battery-Powered Ignition System

A battery-powered ignition system is more common in larger small engines, such as those in lawn mowers, snowblowers, and generators. In this system, the engine has a battery that provides power to the ignition coil. This system generally provides a stronger and more reliable spark than magneto systems, especially at higher engine speeds. Battery-powered ignition systems typically have additional components such as a rectifier and a voltage regulator to manage the power being supplied to the system and ensure consistent ignition.

Key Components of the Ignition System

Ignition Coil: The ignition coil is responsible for converting low voltage from the magneto or battery into the high voltage needed to create a spark at the spark plug.

Spark Plug: The spark plug delivers the electrical spark that ignites the fuel mixture. The spark plug's condition directly affects the engine's ability to start and run smoothly.

Flywheel: In magneto systems, the flywheel is a critical component that spins to generate the electrical current for the ignition system.

Kill Switch: The kill switch (or stop switch) is used to interrupt the flow of electricity to the ignition coil, stopping the engine when needed.

Diagnosing and Replacing Spark Plugs and Ignition Coils

Faulty spark plugs and ignition coils are among the most common reasons for engine misfires or failure to start. Here's a guide on how to diagnose and replace these components.

Diagnosing the Spark Plug

The spark plug is often the first place to check if your engine isn't starting or is running rough. A spark plug can become fouled, worn, or damaged over time, leading to poor engine performance.

Common Symptoms of a Bad Spark Plug:

- Engine won't start or has trouble starting
- Rough idling or misfires during operation
- Poor acceleration or power loss
- Black smoke or excess fuel consumption

STEPS TO DIAGNOSE A SPARK PLUG

Inspect the Spark Plug

First, remove the spark plug using a spark plug socket wrench. Visually inspect it for damage or wear. A spark plug that is clean and dry indicates a possible issue with the ignition system or air-fuel mixture. A spark plug with carbon buildup or oil residue may be a sign of engine problems, such as a rich fuel mixture, worn piston rings, or valve seals.

Check for Spark

After removing the spark plug, reconnect it to the spark plug wire, then ground it against the engine's metal surface. Crank the engine and observe whether a strong spark is visible at the tip of the spark plug. If there is no spark or the spark is weak, the ignition coil or wiring may be faulty.

Test the Gap

Use a spark plug gap tool to measure the gap between the spark plug electrodes. The gap should be set according to the manufacturer's specifications. If the gap is too wide or too narrow, the spark plug may not provide the proper spark needed for combustion.

Replace the Spark Plug

If the spark plug is damaged, corroded, or worn beyond repair, replace it with a new one that matches the engine's specifications. When installing a new spark plug, use a torque wrench to tighten it to the manufacturer's recommended torque value.

Diagnosing the Ignition Coil

The ignition coil is responsible for converting low-voltage power into the high-voltage spark that is needed to ignite the fuel mixture. A malfunctioning ignition coil can lead to a weak or intermittent spark, which will cause the engine to run poorly or fail to start.

Common Symptoms of a Faulty Ignition Coil:

- Engine won't start
- Engine stalls after running for a while
- Rough idling or poor performance
- Intermittent or weak spark

STEPS TO DIAGNOSE THE IGNITION COIL

Inspect the Coil for Damage

Visually inspect the ignition coil for signs of cracks, burns, or other damage. If the coil has visible damage, it may need to be replaced.

Test the Coil With a Multimeter

Use a multimeter to check the resistance of the ignition coil. A typical coil should have a resistance range specified by the manufacturer, usually in the range of 3 to 6 ohms for the primary side and around 5,000 to 10,000 ohms for the secondary side. If the resistance is out of range, the coil may be faulty. When testing the coil, if the multimeter reads zero or has continuity or is grounded out, the coil will need to be replaced.

Test for Spark

After testing the coil with a multimeter, test for a spark. Remove the spark plug, attach the spark plug wire to the plug, and ground the plug against the engine. Crank the engine and check for a spark. A weak or absent spark indicates a problem with the ignition coil.

Replace the Ignition Coil

If the ignition coil is faulty, replace it with a new one. Be sure to install the new coil exactly as the old one was positioned and ensure that all wires are securely connected. When installing a new ignition coil, there needs to be an "air gap" between the coil and the flywheel. A business card provides the correct air gap.

Once the coil is loosely installed, turn the flywheel until it is magnetized on the coil. Tighten the coil to the correct torque and then check for a spark.

Understanding and Testing Kill Switches and Wiring

The kill switch is a critical component in the electrical system, allowing the user to stop the engine by interrupting the ignition system. The kill switch essentially acts as a short to ground, preventing the ignition coil from firing by cutting off the electrical signal to the coil. If the kill switch malfunctions, the engine may not turn off or fail to start.

Understanding the Kill Switch

The kill switch, also known as the stop switch, can be either a manual or automatic switch, depending on the engine. Most small engines, such as lawn mowers or chain saws, are equipped with a manual kill switch that can be activated by pressing a button or pulling a lever. The switch is designed to stop the engine immediately by grounding the ignition coil.

Common Kill Switch Problems:

- Kill switch may fail to stop the engine
- Engine may not start if the kill switch is malfunctioning

Testing the Kill Switch

Inspect the Wiring

Check the wiring connected to the kill switch for any loose connections or signs of wear. Damaged wires could prevent the switch from working properly.

Test the Switch

With a multimeter, check the continuity of the kill switch. When the switch is in the “on” or “run” position, it should show continuity. When the switch is in the “off” or “kill” position, it should not show continuity.

Bypass the Kill Switch

If you suspect that the kill switch is malfunctioning and preventing the engine from starting, you can temporarily bypass the switch by connecting the wire from the kill switch directly to ground. This process will confirm if the kill switch is the problem.

Testing and Repairing Wiring

Wiring issues can cause a variety of electrical problems, including difficulty starting the engine, loss of spark, or intermittent operation.

STEPS TO TEST AND REPAIR WIRING

Inspect the Wires

Visually inspect the wiring for any visible damage, such as fraying, cracks, or burnt areas. Pay special attention to the areas around connectors, terminals, and the ignition coil.

Test for Continuity

Use a multimeter to check for continuity in the wires. Any wire that has no continuity should be replaced.

Repair or Replace Wires

If you find any damaged or broken wires, they should be repaired or replaced. When splicing wires, ensure that you use the correct type of connector and make a clean, secure connection to prevent future issues.

The electrical system in small engines is vital for ensuring proper ignition and smooth operation. Diagnosing and repairing issues with the ignition coil, spark plugs, kill switch, and wiring are essential skills for any small engine owner or mechanic. Regular maintenance and testing of these components can prevent costly repairs and extend the life of your engine. With the right tools and knowledge, you can troubleshoot and fix common electrical problems, keeping your engine running reliably and efficiently.

CHAPTER 10

MECHANICAL REPAIRS

Small engines are made up of several mechanical components that work together to produce power and keep the engine running smoothly. Over time, wear and tear can take a toll on these components, leading to performance issues or total engine failure. This chapter focuses on how to perform essential mechanical repairs, such as replacing worn-out pistons, valves, and gaskets, diagnosing and repairing compression issues, and servicing the crankshaft and camshaft.

Replacing Worn-Out Pistons, Valves, and Gaskets

The piston, valves, and gaskets are critical components in the engine's operation. Each part must work together to ensure proper compression, combustion, and exhaust flow. When any of these components wear out, it can lead to decreased engine performance, rough operation, or complete engine failure

Replacing Pistons

The piston moves up and down within the cylinder, creating the compression needed for combustion and transmitting power from the combustion chamber to the crankshaft. Pistons can wear out over time because of friction, heat, and the constant expansion and contraction that occurs during engine operation. Worn pistons can lead to reduced compression, misfires, or a loss of power.

Signs of Worn Pistons:

- Loss of power or poor acceleration
- Increased oil consumption
- Blue smoke from the exhaust, which indicates oil burning in the combustion chamber
- Low compression readings when testing the engine

STEPS TO REPLACE A WORN PISTON

Remove the Cylinder Head

To access the piston, you will need to remove the cylinder head. This step involves disconnecting any attached components, such as the spark plug, air filter, and fuel lines, before unscrewing the bolts that hold the cylinder head in place. Be careful not to damage the gasket as you remove the head.

Remove the Cylinder

Once the cylinder head is removed, use a socket wrench to remove the cylinder bolts. Carefully lift the cylinder off the engine, revealing the piston inside.

Remove the Old Piston

To remove the piston, you will need to disconnect the connecting rod from the crankshaft. This step can usually be done by removing the rod bolt and sliding the piston out. Inspect the piston and the connecting rod for wear or damage.

Install the New Piston

Install the new piston by carefully aligning it with the connecting rod and sliding it into place in the cylinder. Pay close attention to the top of the piston; if there is an arrow or some sort of indentation on one side of the piston, that arrow or indentation aligns with the exhaust side of the cylinder. It is important to read the instructions that are part of the piston package. If there are no markings, it generally means that the piston can be installed without directions. Reattach the connecting rod to the crankshaft, making sure to torque the rod bolt to the manufacturer's specifications.

Reassemble the Engine

After replacing the piston, carefully reassemble the engine by reinstalling the cylinder, cylinder head, and all other removed components. Be sure to replace the cylinder head gasket to prevent leaks.

Replacing Valves

The intake and exhaust valves control the flow of the air-fuel mixture into the combustion chamber and the expulsion of exhaust gases. Over time, valves can become worn or damaged, leading to issues such as poor compression, misfires, or inefficient combustion.

Signs of Worn Valves:

- Loss of power or performance

- Rough idle or engine misfires
- Difficulty starting the engine
- Low or uneven compression readings

STEPS TO REPLACE WORN VALVES

Remove the Cylinder Head

As with piston replacement, the cylinder head needs to be removed to access the valves. Unscrew the bolts by securing the cylinder head to the engine block and carefully lifting it off, being mindful of the valve springs and components inside.

Remove the Valve Springs and Keepers

Use a valve spring compressor tool to remove the valve springs. This tool compresses the springs so the valve keepers can be removed. Once the keepers are off, you can take out the springs and the valves.

Inspect the Valve Seats

Before installing new valves, check the valve seats (the areas where the valves make contact with the cylinder head) for wear. If the seats are damaged, they may need to be resurfaced using a valve grinding tool.

Install New Valves

Install the new valves by carefully aligning them with their respective valve seats. Reinstall the valve springs, making sure to compress them enough to reattach the keepers securely.

Reassemble the Engine

After the valves are replaced, reassemble the engine by reinstalling the cylinder head, tightening the bolts to the correct torque specification, and replacing any gaskets. Reattach all other components, such as the spark plug and intake manifold.

Replacing Gaskets

Gaskets create seals between the different engine components, preventing air, fuel, or oil from leaking into or out of the engine. Over time, gaskets can degrade because of heat, pressure, and age, resulting in oil or coolant leaks, loss of compression, or other issues.

Common Types of Gaskets

Cylinder Head Gasket: Seals the interface between the cylinder head and the engine block

Base Gasket: Seals the base of the cylinder to the engine block

Exhaust Gasket: Seals the exhaust manifold to the cylinder head

Intake Gasket: Seals the intake manifold to the cylinder head

Signs of Worn Gaskets:

- Oil or coolant leaks around engine components
- White smoke from the exhaust (indicating coolant leaking into the combustion chamber)
- Loss of power or rough running due to loss of compression

STEPS TO REPLACE GASKETS

Identify the Leak

Before replacing the gasket, identify where the leak is coming from. Common areas to check include the cylinder head, exhaust manifold, and intake manifold.

Remove the Old Gasket

Once the leak is identified, remove any components that are obstructing access to the gasket, such as the cylinder head or intake manifold. Carefully remove the old gasket, making sure not to damage the mating surfaces of the engine parts.

Clean the Mating Surfaces

Before installing a new gasket, clean the mating surfaces with a gasket scraper or solvent to remove any old gasket material, carbon buildup, or oil residue. A clean, smooth surface ensures a proper seal.

Install the New Gasket

Place the new gasket in position, ensuring it is properly aligned with the holes in the engine components. Be sure to use the correct type of gasket for your specific engine model because different engines may use different materials or gasket thicknesses.

Reassemble the Engine

Once the gasket is installed, reassemble the engine by reinstalling any removed components, such as the cylinder head, intake manifold, or exhaust system. Tighten the bolts to the manufacturer's torque specifications.

Diagnosing and Repairing Compression Issues

Compression is vital for an engine's performance because it ensures that the air-fuel mixture is properly compressed before ignition. Low compression can result in poor engine performance, difficulty starting, or even engine failure. Compression issues can be caused by a variety of factors, including worn pistons, valves, gaskets, or rings.

STEPS TO DIAGNOSE COMPRESSION ISSUES

Perform a Compression Test

To measure the engine's compression, use a compression tester. Remove the spark plug, screw the tester into the spark plug hole, and crank the engine. The compression tester will display the pressure in the cylinder, and you can compare it to the manufacturer's recommended compression values.

Check the Results

If the compression is low, it could indicate worn piston rings, damaged valves, or blown gaskets. If the compression is significantly different between cylinders, it could also point to uneven wear or internal damage.

Perform a Leak-Down Test

If the compression test reveals low compression, a leak-down test can help pinpoint the source of the leak. This process involves pressurizing the cylinder with compressed air and listening for air escaping through the exhaust, intake, or crankcase. When performing a leak-down test, it is important to not over-compress the engine; this situation can cause major issues to older engines. Typically, 15 to 20 pounds is sufficient enough to pressurize an engine. Once pressurized, wait five minutes for the results. The engine shouldn't lose any more than 10% of the initial pressure that is placed in the cylinder.

STEPS TO REPAIR COMPRESSION ISSUES

Check the Valves

If low compression is caused by worn valves, they will need to be replaced, as described earlier in this chapter.

Inspect the Piston Rings

If the compression is low because of worn piston rings, you will need to replace the rings and possibly the pistons, depending on the degree of wear.

Replace Gaskets

If a blown head gasket is the cause of the compression loss, replacing the gasket should restore proper compression.

Servicing the Crankshaft and Camshaft

The crankshaft and camshaft are integral to the engine's operation. The crankshaft converts the up-and-down motion of the piston into rotational motion, while the camshaft controls the opening and closing of the engine's valves. If these components are damaged or not properly maintained, they can cause significant engine problems.

Inspecting and Servicing the Crankshaft

Symptoms of Crankshaft Issues: Vibration, knocking noises, or difficulty turning over the engine

Steps: Remove the engine block to inspect the crankshaft for wear or damage. Check the bearings and replace them, if necessary. If the crankshaft is bent or cracked, it will need to be replaced.

Inspecting and Servicing the Camshaft

Symptoms of Camshaft Issues: Loss of power, poor valve timing, or misfires

Steps: Remove the cylinder head to access the camshaft. Inspect the camshaft lobes and bearings for wear. Replace any damaged components. If the camshaft is severely worn, it may need to be replaced entirely.

Mechanical repairs on small engines, such as replacing worn pistons, valves, and gaskets, diagnosing and fixing compression issues, and servicing the crankshaft and camshaft, are essential skills for anyone working on engine maintenance. Regular attention to these critical components will ensure the longevity and reliability of your small engine, allowing it to run efficiently for years to come. Proper diagnosis and repair techniques will not only help restore performance but also prevent more severe engine damage in the future.

PART 3

REBUILDING THE SMALL ENGINE

CHAPTER 11

SMALL ENGINE REBUILDS

Rebuilding a small engine is a comprehensive process that requires careful inspection, skillful disassembly, and precise reassembly of critical engine components. For those working with small engines, such as those in lawn mowers, generators, and chain saws, knowing when to rebuild versus replace the engine is a critical decision. This chapter will explore how to make that decision, guide you through the process of disassembling and inspecting engine components, and provide a step-by-step guide to rebuilding a small engine.

DECIDING WHEN TO REBUILD VS. REPLACE AN ENGINE

One of the most common questions small engine owners face is whether to repair or replace their engine. While a brand-new engine might seem like the easiest solution, it may not always be the most cost-effective choice, especially if your current engine is in good overall condition and just requires some restoration.

WHEN TO REBUILD

A rebuild may be the best choice in the following situations.

High-Quality Engine With Good Core Parts: If the engine is from a reputable brand and has a solid history of performance, a rebuild can extend its life at a fraction of the cost of a replacement engine.

Minor Wear or Damage: If the engine has worn components, such as valves, piston rings, or gaskets, but the core components, such as the crankshaft or cylinder head, are in good condition, rebuilding can be a more economical solution.

Sentimental or Practical Value: In some cases, the engine may have sentimental value or be a critical part of equipment that's difficult to replace. For example, vintage lawn mowers or specialty machines may require a rebuild to restore them to their original working condition.

Cost-Effectiveness: If you can perform the engine rebuild yourself or have access to reasonably priced parts and professional assistance, rebuilding the engine can save you money, especially if the cost of a new engine is high.

WHEN TO REPLACE

In contrast, you may want to replace the engine when the following happens.

Severe Damage to Core Components: If critical parts such as the crankshaft, camshaft, or engine block are cracked or severely damaged, replacing the engine is likely more economical than rebuilding.

Outdated or Irreparable Engines: Older engines may be so outdated that replacement parts are hard to find or prohibitively expensive. In such cases, purchasing a new engine could be the best solution.

Frequent Repairs: If an engine has required multiple repairs over a short period, or if it has a history of chronic problems, it may be more cost-effective to replace it entirely rather than keep repairing it.

High Replacement Costs: In some situations, the cost of a rebuild (e.g., parts, labor, and time) can approach or exceed the price of a new engine. If a new engine comes with a warranty and longer life expectancy, it may make more sense to replace it.

Making this decision ultimately comes down to the condition of the engine, the cost of repair parts, and how much time and money you're willing to invest in restoring it.

DISASSEMBLING AND INSPECTING ENGINE COMPONENTS

Once you've decided to rebuild your engine, the next step is disassembly. This step is a critical phase of the rebuilding process because it allows you to assess the condition of individual components. Proper disassembly requires patience, precision, and organization.

Tools You'll Need for Disassembly:

- Socket wrenches and ratchets (with a variety of sockets)
- Torque wrenches
- Screwdrivers (both flathead and Phillips)
- Pliers and needle-nose pliers
- Gasket scraper
- Magnetic tray or parts bin for keeping screws and small components organized
- Engine hoist (for larger engines)

STEPS TO DISASSEMBLE THE ENGINE

Prepare the Workspace: Ensure you have a clean, well-lit workspace with enough room to lay out engine parts as you remove them. Having an organized system for storing parts will save you time later when reassembling the engine.

Disconnect the Battery (if applicable): If the engine is part of equipment that uses a battery (e.g., a lawn mower or generator), disconnect the battery to prevent electrical shorts during the disassembly process.

Remove the Engine From the Equipment: Depending on the equipment, you may need to remove the engine from its mounting or housing. Use appropriate lifting equipment to handle the engine safely, especially if it's heavy.

Drain Fluids: Drain the engine of any remaining fuel, oil, or coolant. This step will make the disassembly process cleaner and safer.

Remove External Components: Begin by removing any external components, such as the air filter, exhaust system, carburetor, fuel lines, and ignition system. Keep track of screws and bolts in labeled containers to ensure easy reassembly.

Remove the Cylinder Head: Unscrew the bolts that secure the cylinder head and carefully lift it off the engine. Inspect the cylinder head for cracks or signs of warping. Also, check the condition of the valves and valve springs. If the valves appear worn, they may need to be replaced.

Remove the Piston and Connecting Rod: After removing the cylinder head, you can access the piston. Remove the piston by detaching the connecting rod from the crankshaft. Carefully slide the piston out of the cylinder and inspect it for scoring or cracks. The piston rings should also be checked for wear or breakage.

Remove the Crankshaft: Finally, remove the crankcase and any remaining components. This step will give you access to the crankshaft, which can be checked for bending, scoring, or wear on the bearings. A damaged crankshaft is often a reason for replacing the engine rather than rebuilding it.

Inspect Each Component: After disassembling the engine, carefully inspect each component for damage or excessive wear. Critical components to inspect include:

- Cylinder walls for scoring or excessive wear
- Piston rings for wear or cracks
- Connecting rod and bearings for signs of wear or play
- Crankshaft for cracks, bending, or wear on the journals

- Valve seats and valves for damage or pitting
- Gaskets to see if they've deteriorated and need replacement

Now that you've disassembled and inspected the engine, the next step is to rebuild it. A successful rebuild involves replacing worn parts, cleaning components, and ensuring that everything is reassembled correctly to restore engine performance.

STEPS TO REBUILD THE ENGINE

Clean and Prepare Components: Before reassembling the engine, clean all components thoroughly. Use a parts washer or solvent to remove oil, dirt, and carbon buildup. Inspect and clean the carburetor, cylinder head, and other components that can be reused.

Replace Worn or Damaged Parts: Replace any parts that are excessively worn or damaged. This step includes piston rings, valves, gaskets, seals, and bearings. Ensure that replacement parts are of high quality and match the engine's specifications.

Reassemble the Crankshaft: Start by reinstalling the crankshaft, ensuring that it is properly aligned and supported by bearings. Torque the crankshaft bolts to the manufacturer's specifications.

Install the Piston and Connecting Rod: Slide the new piston rings onto the piston (if replacing). Carefully reinsert the piston into the cylinder, aligning the piston with the connecting rod. Attach the connecting rod to the crankshaft and torque the bolts securely.

Install the Cylinder Head: After the piston and connecting rod are in place, reinstall the cylinder head. Make sure to use a new head gasket to ensure a proper seal. Tighten the cylinder head bolts in a crisscross pattern to avoid warping the head.

Install the Valves: If the valves were replaced, install them into the cylinder head and secure them with new valve springs and keepers. Make sure the valves are seated properly against the valve seats. If needed, adjust the valve lash to the correct specification.

Reinstall External Components: Reinstall the carburetor, air filter, fuel lines, ignition system, and exhaust components. Ensure that all connections are tight and secure, then replace any worn or cracked hoses and cables.

Lubricate Moving Parts: Before closing up the engine, lubricate all moving parts, including the piston, crankshaft, camshaft, and bearings. This step helps prevent premature wear when the engine is started for the first time after the rebuild.

Torque the Bolts: Go through all of the bolts and fasteners in the engine and torque them to the manufacturer's specifications. This step ensures that the engine is reassembled correctly and avoids problems with loose parts during operation.

Check Clearances and Timing: Make sure that all clearances (e.g., valve lash, piston-to-wall clearance) are within the manufacturer's specifications. If your engine has a timing system (e.g., camshaft timing), ensure that the timing marks align correctly.

Fill With Fluids: Refill the engine with the correct type and amount of oil, fuel, and any other necessary fluids (e.g., coolant, if applicable).

Test Run: Once the engine is fully reassembled, conduct a test run. Monitor for any abnormal noises, vibrations, or leaks. Check that the engine starts easily and runs smoothly. If there are any issues, stop the engine immediately and troubleshoot.

Rebuilding a small engine is an intricate but rewarding process that can extend the life of an engine and save you money. By knowing when to rebuild versus replace, carefully disassembling and inspecting engine components, and following a detailed step-by-step guide, you can successfully restore an engine to like-new condition. With patience, the right tools, and a focus on precision, you'll be able to give your small engine a new lease on life, improving its performance and ensuring its continued operation for years to come.

PART 4

PREVENTION

CHAPTER 12

PREVENTING FUTURE ISSUES

Preventing future issues in small engines involves not only proper maintenance but also adopting the best practices for long-term care, storage, and environmentally responsible disposal. Small engines, such as those in lawn mowers, chain saws, generators, and outboard motors, are built to last, but without proper upkeep, they can suffer from performance loss, early wear, and even failure. By developing a comprehensive approach to maintenance that includes regular seasonal checks, long-term care strategies, and responsible disposal of old parts and fluids, you can extend the life of your engine and keep it running efficiently year-round.

Long-Term Care and Storage Tips

Proper long-term care and storage are essential to prevent unnecessary damage and ensure that your small engine is ready to perform when needed. Whether you're storing an engine for the offseason or simply maintaining it for long-term use, taking the right steps can preserve its functionality and prevent costly repairs.

Clean Before Storing

Before storing your engine for an extended period, it's essential to clean it thoroughly. Dirt, debris, and fuel residue can damage internal components and reduce the lifespan of the engine. Here's how to properly clean your engine.

Exterior Cleaning: Remove any dirt, grass, leaves, or other debris from the engine's exterior. Use a soft brush, compressed air, or a damp cloth to clean the outer surfaces. Be careful not to spray water directly into sensitive areas such as the carburetor or ignition system.

Fuel System Cleaning: Drain any remaining fuel from the engine to prevent it from degrading and causing clogging or corrosion in the carburetor and fuel lines. Use a fuel stabilizer if you're storing the engine with fuel in the tank. This step helps prevent the fuel from breaking down and gumming up the system.

Air Filter Cleaning: Check and clean the air filter. If it's clogged or too dirty, replace it before storage. A clean air filter ensures optimal air intake when the engine is started again.

Protect From the Elements

If you're storing the engine outdoors or in a garage, protect it from moisture and extreme temperature fluctuations that can cause rust or frozen components. Here's what you can do.

Cover the Engine: Use a protective cover to shield the engine from dust, dirt, and moisture. This step is especially important for engines stored in unsealed areas such as sheds or garages.

Avoid Damp Storage Areas: If possible, store engines in a dry area where they won't be exposed to excess humidity. Damp conditions can lead to rust on exposed metal parts, which can significantly reduce the engine's lifespan.

Consider Elevated Storage: If storing the engine on the ground, elevate it on a platform or shelf to avoid exposure to standing water and moisture.

Run the Engine Regularly

Even if the engine is not in regular use, it's important to start it periodically to keep the internal components lubricated and prevent seals from drying out. Run the engine for 10-15 minutes every 4-6 weeks, especially in the offseason. This practice will help maintain the fuel system's integrity, keep the battery charged (if applicable), and ensure smooth engine operation when you need it again.

Seasonal Maintenance Routines

Small engines often experience periods of heavy use followed by extended downtimes, especially with equipment such as lawn mowers, generators, and snowmobiles. Each season presents unique challenges for small engines, and performing seasonal maintenance is key to preventing breakdowns and ensuring smooth operation.

Spring Maintenance (for equipment such as lawn mowers and trimmers)

After a long winter, small engines used in warmer months (such as lawn mowers, weed trimmers, and garden tools) need a thorough inspection to prepare for the upcoming season. Follow these steps to ensure your equipment is ready for use.

Oil Change: Change the oil to ensure smooth lubrication during the growing season. Old oil can become thick and gritty, reducing engine efficiency.

Inspect the Spark Plug: Remove, inspect, and clean or replace the spark plug, if necessary. A clean spark plug ensures proper ignition and reduces engine misfires.

Check the Fuel System: If you stored the engine with fuel in it, check the fuel system for clogging or gum buildup. Drain any old fuel and refill fresh gasoline. Clean the carburetor, if needed, and replace the fuel filter.

Sharpen the Blades: For lawn mowers, sharpen the mower blade to ensure clean cuts and prevent strain on the engine. A dull blade can lead to more engine wear and inefficient performance.

Check the Air Filter: Ensure the air filter is clean and replace it if it's dirty or clogged. The air filter keeps dirt and debris from entering the engine, helping it run more smoothly.

Fall/Winter Maintenance (for equipment such as snowmobiles and generators)

As the season transitions to colder months, snowmobiles, generators, and other cold-weather equipment require special attention to avoid starting problems and engine damage in freezing temperatures.

Change the Oil: Just like in spring, it's important to change the oil at the end of the season. This step ensures that the engine runs on clean oil during the next use, avoiding corrosion or engine damage from degraded oil.

Fuel Stabilizer: Add a fuel stabilizer to the fuel tank if you plan to store the engine for an extended period. This practice will prevent the fuel from breaking down and clogging the fuel system. If you're storing a snowmobile or outboard motor, run the engine briefly with the stabilized fuel to ensure it gets into the carburetor and fuel lines.

Battery Maintenance: If your engine has a battery, clean the terminals and check the charge. If the battery is not being used, disconnect it or remove it for storage. You can also consider using a battery tender to keep it charged during the off-season.

Inspect Belts and Cables: Check all belts, cables, and hoses for signs of wear or cracking. If any parts appear damaged, replace them to avoid issues when the engine is next used.

Environmental Considerations and Proper Disposal of Old Parts and Fluids

Small engine maintenance and repairs often produce waste, including used oil, old parts, fuel, and other fluids. It's essential to dispose of these items responsibly to minimize environmental impact and comply with local regulations.

Used Oil and Fluids

Used oil is a hazardous material and must be disposed of properly. Pouring it down the drain or into the ground is illegal and harmful to the environment. Here's how to dispose of used oil and fluids correctly.

Take to a Recycling Center: Most auto parts stores, service stations, and local recycling centers accept used motor oil. They will typically collect it for recycling, ensuring it doesn't harm the environment. Check your local regulations for the closest drop-off locations.

Fuel Disposal: If you need to dispose of old or stale fuel, do not pour it into the ground or drain. Several recycling centers also accept old gasoline. Alternatively, you can add the fuel to a car's gas tank in small amounts to burn it off.

Old Parts and Components

Old engine parts such as filters, spark plugs, and belts are usually made from a combination of metals and plastics. These items can often be recycled, but it's important to separate them according to material type.

Scrap Metal: Several parts, such as the metal components of a carburetor or spark plug, can be recycled at scrap metal facilities.

Plastics and Rubber: Items such as air filter housings and belts may not be recyclable in the same way as metals. However, some specialty recycling centers may accept these materials. Check with local recycling facilities for guidance.

Proper Cleaning Supplies Disposal

Cleaning supplies, such as solvents, degreasers, or lubricants, can also pose environmental risks. If you have used solvents or cleaners to wash engine components, dispose of them in accordance with local hazardous waste disposal guidelines. Never pour them down the drain or into the ground.

Preventing future issues with small engines requires a proactive approach that incorporates long-term care, seasonal maintenance, and responsible disposal practices. Regularly servicing your engine, especially as the seasons change, ensures that it continues to operate efficiently and extends its lifespan. Storing engines properly and maintaining key components, such as the fuel system and spark plugs, helps prevent issues when the engine is called into action again. Additionally, environmentally responsible disposal of used fluids and parts is an essential part of maintaining not just the engine but also the surrounding environment.

By adopting these preventive strategies and committing them to regular maintenance routines, you can minimize the likelihood of engine failures, reduce repair costs, and enjoy the reliable performance of your small engine for years to come.

THE PATH FORWARD

Small engine repair is a dynamic and ongoing learning process, where the world of small engines is constantly evolving. As technology advances, new materials, engine designs, and innovative features emerge, making it essential for mechanics to stay current with these changes. While the basic principles of engine repair remain consistent, new engines often come equipped with electronic systems, digital controls, and more complex fuel management features. Keeping your skills up to date will ensure that you remain proficient in diagnosing and fixing modern engines while staying ahead of the curve in the industry.

This book provides a strong foundation in small engine repair, offering essential knowledge that serves as a jumping-off point for further learning. It emphasizes the importance of staying curious and constantly seeking ways to improve your understanding of engine operation. The more experience you gain, the more confident you will become in tackling increasingly complicated repairs and troubleshooting more sophisticated problems.

As your skills develop, you will be able to refine your techniques and approach engine repairs with greater precision. The book encourages hands-on experimentation, which is an invaluable part of becoming a proficient mechanic. By exploring new modifications, customizations, or improvements, you can elevate your repair work and expand your capabilities. Whether you're enhancing engine performance, restoring vintage equipment, or customizing engines for specific uses, your growing expertise will empower you to take on more ambitious projects and explore new opportunities in the world of small engine repair.

FINAL THOUGHTS

"A Basic Guide to Small Engine Repair" stands as a comprehensive and invaluable resource for anyone interested in understanding, maintaining, and repairing small engines. Whether you are a homeowner looking to maintain your lawn mower, a DIY enthusiast eager to expand your mechanical knowledge, or a professional mechanic seeking to deepen your expertise, this book provides the essential tools and information necessary for success in small engine repair. With its well-organized, step-by-step approach, the guide offers a wealth of knowledge, making it accessible to both novices and seasoned mechanics alike.

The book's primary strength lies in its foundational approach to small engine mechanics. It begins by explaining the basic principles of engine operation, giving readers a solid understanding of how small engines function. This foundational knowledge is critical for troubleshooting and performing maintenance because it helps readers easily identify issues and understand the necessary steps to resolve them. Whether you are diagnosing a common problem, such as a sputtering engine, or addressing more complex mechanical failures, having a clear grasp of engine components and the principles of combustion will simplify the process and enable you to take the right actions with confidence.

An important focus of the guide is the emphasis on routine maintenance to extend the life of small engines and prevent costly repairs. The book explains in detail essential tasks such as changing oil, replacing filters, and servicing spark plugs. Regular maintenance not only improves engine performance but also ensures your equipment runs efficiently, saving you money on fuel and reducing the likelihood of serious mechanical breakdowns. By learning these preventive measures, readers can avoid expensive repairs while improving the overall longevity and reliability of their engines.

However, even with the best maintenance practices, problems are inevitable. This reality is where the book's troubleshooting and repair techniques become invaluable. Readers are guided through diagnosing common issues, such as poor starting, rough running, and stalling. With practical diagnostic checklists and step-by-step troubleshooting methods, the guide empowers readers to identify the root causes of engine failures and resolve them independently. This ability to tackle problems on your own not only saves time and money but also fosters self-reliance and confidence, allowing you to address minor issues before they become more significant, costlier repairs.

For those wishing to take their skills to the next level, the guide delves into more complex topics, such as engine rebuilding and working with electrical systems. These advanced sections offer detailed explanations and techniques that are designed to enhance your expertise and deepen your understanding of engine mechanics. Whether restoring vintage engines or repairing modern machinery, mastering these advanced skills will improve your proficiency and open up a broader range of repair opportunities, increasing both your capabilities and confidence.

Safety is a primary concern when working with small engines, and the book doesn't overlook this essential aspect. It emphasizes the importance of using proper safety gear, handling fuel responsibly, and ensuring that workspaces are well ventilated. By following these critical safety guidelines, readers can carry out repair projects effectively and securely, minimizing risks to themselves and others. This focus on safety not only ensures that repair tasks are completed without incident but also helps create a safer environment when working on or around small engines.

The guide also provides valuable insights into different types of small engines, including two-stroke and four-stroke engines, as well as the specific maintenance and repair needs of each Whether you are working on a lawn mower, generator, chain saw, or snowblower, the book offers tailored advice for each type of engine, increasing its versatility and applicability. This information ensures that readers are well equipped to handle a wide variety of small engine equipment, enhancing their repair capabilities and expanding their knowledge across different engine systems.

Small engines are integral to numerous industries, ranging from landscaping and construction to emergency power generation. The skills imparted in this guide open up a range of professional opportunities. Whether you seek to expand your repair services, work on heavy-duty equipment, or enter a new career path in small engine repair, this book provides the knowledge necessary to succeed. By understanding how to repair and maintain small engines, readers can broaden their professional horizons, offering valuable services or working on more complex machinery in a variety of industries.

For hobbyists, the satisfaction of successfully repairing and maintaining small engines is deeply rewarding. The skills learned from this guide provide practical, hands-on knowledge that can last a lifetime. Small engine repair not only offers the chance to solve problems and save money but also brings the joy of mastering a valuable skill. The book presents these skills in an accessible manner, making it an enjoyable and fulfilling endeavor for those passionate about mechanics.

Additionally, the book stresses the importance of maintaining equipment in optimal working conditions. It covers aspects such as proper storage, fuel management, and seasonal preparation, all of which are essential for preventing neglect-related problems and ensuring the longevity of your engines. By following the practical tips in the final chapters, you can protect your small engines from wear and tear, guaranteeing that they continue to perform reliably for years to come.

Whether you are handling simple repairs on your lawn mower or undertaking more complex engine rebuilds, “A Basic Guide to Small Engine Repair” offers all the guidance you need to succeed. With its clear instructions, expert advice, and useful troubleshooting steps, the book allows you to approach small engine tasks with confidence. By the time you finish reading, you will have the skills and knowledge to diagnose, maintain, and repair a wide range of small engines, saving both time and money while ensuring your equipment remains in top condition.

More than just a practical tool for DIY enthusiasts, “A Basic Guide to Small Engine Repair” is a comprehensive resource for anyone who owns, operates, or maintains small engines. Whether you’re looking to keep your lawn equipment in working order or deepen your mechanical craft as a professional, this book offers the expertise and knowledge needed to thrive. It empowers readers to understand the inner workings of small engines, providing them with the tools to maintain their equipment and keep it running smoothly for years to come.

BIBLIOGRAPHY

Honda Small Engines Repair Manual

Editors. (1998). Honda Small Engines Repair Manual. Clymer Publications.

This comprehensive guide provides detailed instructions for repairing and maintaining Honda small engines. It includes step-by-step procedures, troubleshooting tips, and diagrams for a range of Honda small engine models, covering common engine issues and servicing tasks.

Briggs and Stratton Repair Manual

Editors. (2007). Briggs & Stratton Repair Manual. Haynes Publishing.

A thorough manual offering detailed information for repairing and maintaining Briggs & Stratton engines. It includes guidance on diagnosing engine problems, as well as detailed disassembly and reassembly procedures. The manual serves as a valuable resource for both novice and experienced mechanics.

Kohler Small Engine Repair Manual

Editors. (2007). Kohler Small Engine Repair Manual. Clymer Publications.

This manual focuses on Kohler small engines, providing in-depth repair instructions, maintenance advice, and common troubleshooting procedures. With a focus on Kohler engine systems, it offers insights into engine components and how to address issues such as starting problems, carburetor issues, and electrical concerns.

Tecumseh Small Engine Repair Manual

Editors. (2003). Tecumseh Small Engine Repair Manual. Haynes Publishing.

Offering practical advice on how to repair and maintain Tecumseh engines, this manual is designed to assist technicians with servicing and diagnosing various models. It features step-by-step instructions, illustrations, and tips to help users tackle both minor and major engine repairs.

Kawasaki Small Engine Repair Manual

Editors. (2002). Kawasaki Small Engine Repair Manual. Clymer Publications.

The Kawasaki Small Engine Repair Manual provides extensive coverage of Kawasaki small engine models, offering repair, maintenance, and troubleshooting guidance. This manual is tailored to help users understand engine systems and how to address typical engine failure efficiently.

www.ingramcontent.com/pod-product-compliance
Lightning Source LLC
LaVergne TN
LVHW022043200926
848481LV00003B/764